80 Activities to Make Basic Algebra Easier

Second Edition

Robert S. Graflund

J. WESTON
WALCH
PUBLISHER

Portland, Maine

User's Guide
to
Walch Reproducible Books

As part of our general effort to provide educational materials that are as practical and economical as possible, we have designated this publication a "reproducible book." The designation means that purchase of the book includes purchase of the right to limited reproduction of all pages on which this symbol appears:

Here is the basic Walch policy: We grant to individual purchasers of this book the right to make sufficient copies of reproducible pages for use by all students of a single teacher. This permission is limited to a single teacher and does not apply to entire schools or school systems, so institutions purchasing the book should pass the permission on to a single teacher. Copying of the book or its parts for resale is prohibited.

Any questions regarding this policy or requests to purchase further reproduction rights should be addressed to:

Permissions Editor
J. Weston Walch, Publisher
321 Valley Street • P.O. Box 658
Portland, Maine 04104-0658

1   2   3   4   5   6   7   8   9   10

ISBN 0-8251-4261-X

Copyright © 1983, 2001
J. Weston Walch, Publisher
P.O. Box 658 • Portland, Maine 04104-0658
www.walch.com

Printed in the United States of America

Contents

Chapter 7: Using Fractions in Algebra

Chapter 8: Graphing and Systems of Linear Equations

Chapter 9: Rational and Irrational Numbers

Introduction

80 Activities to Make Basic Algebra Easier was developed with these elements in mind:

1. A number of current basic algebra texts were examined for content. This book attempts to include those topics most commonly taught in a first-year algebra course.

2. A manipulative approach to teaching algebra that allows the student to actually sense what is happening can be of great value. However, this approach presents the teacher with several problems.

   (a) Many older students seem to resent the use of various manipulative devices and the teacher will be confronted with remarks like, "That's baby stuff," or "That's what the little kids do."

   (b) Collecting and distributing various manipulative devices to each student can be time-consuming and in some cases expensive.

   In an attempt to overcome these problems, we present a series of activities that require students to make or revise drawings to represent the steps in various algebraic processes. This approach is used a great deal in Chapter 2, "Solving Basic Equations." The student is asked to add (draw in) or subtract (cross out) the same amount from both sides of equations represented by drawings. This method allows the student to become actively involved in the equation-solving process in a mature way, yet it doesn't require any extra materials.

It is our hope that this approach will allow even the slowest student to understand the logic involved in the equation-solving process. Unfortunately, as most teachers know, when you slow the class down in order to help some students, those students that already understand the process quickly become bored with the activity and can become disruptive.

One solution to this problem is to place some of the drawings on the blackboard and have the students who already seem to understand the process show the rest of the class what to do. This helps to inflate the good student's ego and allows the slow student to *see* the logic that is involved. If an overhead projector is available, physical objects can actually be added or subtracted (removed) from both sides of the equation. A somewhat theatrical approach can be provided by using small cardboard boxes to represent variables and placing the number of objects that represent the answer inside the box. After the equation has been solved by manipulating the objects, the box can be opened to reveal the answer. Please note that these activities should be introduced to the students in the same logical order in which they appear in the book.

3. Some of the concepts in this book are developed by relating the algebraic concepts to what the students have done in arithmetic. If you consider that prior to taking algebra the students have spent most of their time in math class doing arithmetic,

then it should not be surprising that they will more *readily* accept an idea in algebra if it can be related to an example from arithmetic.

4. In some cases students are led to discover an idea through the use of patterns. In this inductive approach, the student is required to solve a set of simple problems and is then asked a series of questions involving those problems. This process allows the student to discover the desired concept gradually.

5. Once the concept has been developed it is further reinforced through the use of puzzles (Dot-to-Dot Puzzles, Magic Squares, Cross-Number Puzzles, Codes, etc.). For the most part these puzzles are self-correcting and allow students to know quickly if they are proceeding in the right direction. Puzzles tend to relax the over-anxious student and also to provide a change of pace from traditional exercises.

6. Cartoon figures tend to stress key ideas, and in some cases reflect student attitudes toward math. (Seeing their own negative thoughts in print also helps students to relate to the book.)

7. Some of the activities in the book lend themselves to group work. Examples would be the drawing used to solve equations, as well as the outdoor activities involving 3-4-5 triangles and similar triangles.

8. The teaching guide page for each chapter gives a brief overview of concepts, activity topics, and a list of materials and prerequisite activities that may be required in the chapter.

This book is intended to be used by students in a first-year algebra course, but there is no reason why it couldn't be used as enrichment for good students in the lower grades.

CHAPTER 1

Getting Started

This chapter will introduce or review some of the basic concepts students will need in order to solve more advanced problems in algebra.

Teaching Tips

- How many times have you had a student tell you that $3^2 = 6$ instead of $3 \times 3 = 9$? Graph-paper drawings and cube models (see Activity 2) can help students visualize this concept.

- Even mature math students hate showing the steps they have followed in solving equations, and having to give reasons (properties) for those steps ("Why must I prove something that's so obvious?"). Because of this, it is suggested that you introduce basic properties early in the program as having value in mental-math shortcuts. Properties should be reintroduced later when students are ready for a more sophisticated approach to solving equations.

Overview of Activities in this Chapter

1. **Order of Operations**

   Overview: Gives students a brief review of the order of operations, with a cross-number puzzle practice activity.

2. **Basic Exponents I**

   Overview: Offers students a visual approach to representing multiplication and exponents.

3. **Basic Exponents II**

   Prerequisite Activity: Activity 2

   Materials: 30 identical cubes (plastic, wood, etc.) for each student or group

   Overview: Builds on prior student work to represent multiplication with three factors, using cubes as manipulatives.

1

Chapter 1: Getting Started

4. **Exponents and Calculators**

   Materials: Scientific calculators

   Overview: Introduces using a scientific calculator to find exponents.

5. **Algebraic Expressions**

   Overview: Introduces some of the essential vocabulary of algebra: *variable*, *term*, and *algebraic expression*.

6. **Something of Interest**

   Overview: Uses a real-life example—interest on savings—to introduce the idea of variables in a formula.

7. **The Compound Interest Formula**

   Overview: Again using a real-life banking example, introduces the use of exponents in formulas.

8. **Formula Review**

   Overview: Reviews students' knowledge of essential geometric formulas through a self-checking dot-to-dot puzzle.

9. **Basic Properties**

   Overview: Reviews the seven basic properties essential to algebra.

B. PEMBAS 5(3)

5(3+5)

ACTIVITY 1

Order of Operations

How would you solve this problem?

$5 + 2 \cdot 3$ (The dot means "multiply.")

6
9/11
8

7×21

Easy!
$5 + 2 = 7$
$7 \times 3 = 21$

But, it could be
$5 + (2 \times 3) =$
$5 + 6 = 11$

Which answer is correct? In order to avoid confusion, mathematicians have come up with the following rules:

1. First, do all the multiplication and division in order from left to right.

2. Next, do any addition and subtraction in order from left to right.

3. If there are parentheses, do what is inside the parentheses first. Then continue with rules 1 and 2.

7 × 3

21

21

If we follow the rules, the answer to $5 + 2 \cdot 3$ is 11.

I knew that I was right.

Who asked you?

Directions Solve the following. (Check your work with your teacher before you go on.)

1. $5 \cdot 3 + 4 =$ _____

2. $5 \cdot (3 + 4) =$ _____

3. $3 + 2 \cdot 6 =$ _____

4. $(3 + 2) \cdot 6 =$ _____

5. $3 \cdot 9 - 5 =$ _____

6. $12 - 2 \cdot 6 =$ _____

7. $3 \cdot 2 + 9 \div 3 =$ _____

8. $12 - 6 \div 3 =$ _____

(continued)

3   *80 Activities to Make Basic Algebra Easier*

Order of Operations *(continued)*

Name _____

Date _____

Directions   Solve each of the numbered math problems below. Then, use the answers to solve the cross-number puzzle.

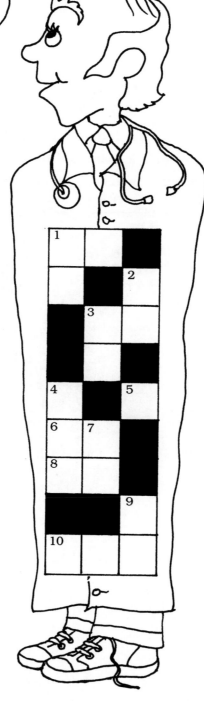

Across

1. $3 \cdot 5 + 6$

3. $3 \cdot (5 + 6)$

5. $3 + 4 \div 2$

6. $2(200 + 76) \div 6$

8. $76 - 6 \cdot 0$

10. $\dfrac{325 - 10}{4 - 1}$

Down

1. $2 \cdot 15 - 2$

2. $19 - 2 \cdot 3$

3. $3(15 - 2)$

4. $5 + 96 \cdot 2$

7. $5 \cdot 5 + 6 \div 6$

9. $[9 - (12 \div 3)] \cdot 5$

Name _____

Date _____

Basic Exponents I

Multiplication with whole numbers can be represented using rectangles.

Example 1:

$3 \times 4 =$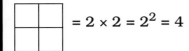

> Just count the squares.

A special kind of problem occurs when we use **squares**.

Example 2:

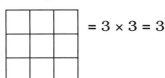 $= 2 \times 2 = 2^2 = 4$ This can be read as "two to the second power" or "two squared."

Example 3:

 $= 3 \times 3 = 3^2 = 9$ The "2" is called the **exponent** and the "3" is called the **base**. (The exponent tells you how many times to use the base as a factor.)

Directions   What exponent problems are shown below? The first problem is done for you. Do the rest the same way.

1. 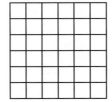 $4^2 = 4 \times 4 = 16$

4^2 $3 \atop 2$

2. 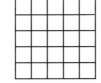   _____

$2 \times 2 \times 2$

3.   _____

(continued)

Basic Exponents I *(continued)*

4. 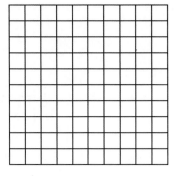 _____

5. 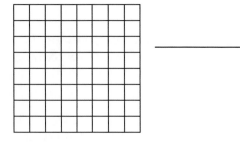 _____

Directions   Use multiplication to complete the following:

6. $9^2 =$ _____

7. $12^2 =$ _____

8. $7^2 =$ _____

9. $1^2 =$ _____

10. Make two drawings that show the difference between 4×2 and 4^2. (***Hint:*** One drawing will be a rectangle. The other drawing will be a square.)

Name _____

Date _____

Basic Exponents II

| **Materials** |
| 30 plastic or wooden cubes |

You have learned how multiplication problems with two factors can be represented using rectangles. Now we will represent multiplication with three factors using rectangular solids made up of cubes.

Example 1: Represent $2 \times 3 \times 5 = 30$

Your model should look like this:

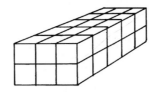

Example 2: Represent $2 \times 2 \times 2 = 8$

$2 \times 2 \times 2$ can be written as 2^3.
(This can be read as "2 to the third power" or "2 cubed.")

Directions   Solve the following by making a model using cubes and counting the cubes. Check your work using multiplication. See if you can make a sketch of each model next to the problem.

1. $2 \times 3 \times 4$ 2. 3^3 3. 3×2^2

If $3^2 = 3 \times 3 = 9$, and $3^3 = 3 \times 3 \times 3 = 27$, what would $3^4 = $ ____?

Directions   Use multiplication to solve the following:

4. $2^4 = $ _____   6. $3^5 = $ _____   8. $10^3 = $ _____

5. $2^5 = $ _____   7. $6^3 = $ _____   9. $1^8 = $ _____

10. Make two drawings that show the difference between 2×3 and 2^3.

Name _____

Date _____

Exponents and Calculators

To find 2^3 on a scientific calculator, we could simply multiply $2 \times 2 \times 2$. However, this could become extremely difficult and time-consuming in some exponent problems. To work with exponents on a calculator, locate a button that looks like this: $\boxed{y^x}$

To find 2^3, press 2 $\boxed{y^x}$ 3 =. The answer should be 8. Some calculators may have x^y or other letters. You may have to use the INV or SHIFT key to activate the exponents function.

Directions   Try the following problems on your calculator:

1. 4^3 = _____   3. 12^3 = _____   5. 6.34×10^3 = _____

2. 2^5 = _____   4. $(3.4)^3$ = _____

Calculator manufacturers have included a special key for squaring numbers.

Example: Find 5^2.

Solution: Enter 5, then press x^2.   Answer: 25.

Directions   Use what you learned about calculators to solve the following problems:

6. 4^2 = _____   8. 100^2 = _____   10. 4^0 = _____

7. 8^2 = _____   9. $(4.2)^2$ = _____

Did the answer to problem 10 surprise you? _____

Directions   Try the following problems on your calculator. Look for a pattern in your answers.

11. 2^4 = _____   16. 5^4 = _____

12. 2^3 = _____   17. 5^3 = _____

13. 2^2 = _____   18. 5^2 = _____

14. 2^1 = _____   19. 5^1 = _____

15. 2^0 = _____   20. 5^0 = _____

What do you think "zero power" means? _____

_____

_____

ACTIVITY 5

Algebraic Expressions

In algebra, a letter that represents a number is called a **variable**.

Multiplication can be shown by placing variables next to each other.
Example: acy means $a \times c \times y$

You can also place a numeral to the *left* of a variable to show multiplication.
Example: $3y$ means "three times y."

A **term** can be a number, a variable, or the product or quotient of numbers and variables. The following are examples of terms:

$6, \ y, \ ab, \ \dfrac{3}{b}, \ \dfrac{a}{2y}, \ 6a$

Connecting one or more terms by addition or subtraction forms an **algebraic expression**.

The following examples show how to "evaluate" an expression.

Example 1:	If a is 2 and c is 3, find the value of $ac + 5$. $(2 \times 3) + 5 = 6 + 5 = 11$ (substitute 2 for a, and 3 for c)
Example 2:	If x is 3, find the value of $2x^2$: $2 \cdot 3^2 = 2 \times 3 \times 3 = 2 \times 9 = 18$
Example 3:	If x is 3, find the value of $(2x)^2$ $(2 \cdot 3)^2 = (2 \cdot 3) \cdot (2 \cdot 3) = 6 \cdot 6 = 36$ $2x^2 = 2 \cdot x \cdot x$ $(2x)^2 = 2x \cdot 2x$ or $4x^2$

(continued)

Name _____

Date _____

Algebraic Expressions *(continued)*

Directions A daring young math student, Vara Variable, has just made contact with some visitors from outer space. Unfortunately, she can read only part of their message. Help her out by solving the numbered problems below the spaceship. For example, solve problem #1. Then locate the answer to problem #1 in the Answer Box. Write the letter that is above that answer in the square marked "1" in the spaceship. Continue in the same way for problems 2–22. If your work is correct, the rest of the message will appear.

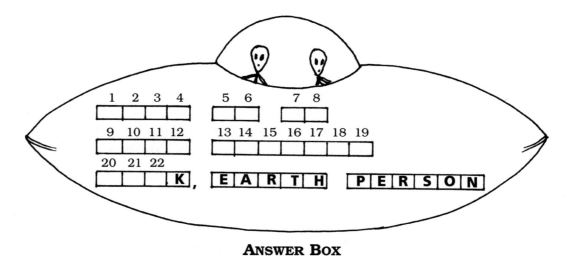

ANSWER BOX

A	B	E	G	K	L	M	O	R	T	U	Y
9	25	10	27	24	0	1	14	12	8	5	13

Evaluate the following where $a = 2$, $b = 3$, $c = 4$:

1. ac _____

2. $a + b + c$ _____

3. abc _____

4. $ab + c$ _____

5. $a + b - c$ _____

6. $bc - a$ _____

7. a^3 _____

8. $a(b + c)$ _____

9. $b^2 + c$ _____

10. $5a + c$ _____

11. $c - a + b$ _____

12. $3a + 2b$ _____

13. $2b + 3$ _____

14. $bc - 6a$ _____

15. $3b^2$ _____

16. $3c - a$ _____

17. $b^2 + c^2$ _____

18. $\dfrac{6c}{a}$ _____

19. $\dfrac{ab^2}{2}$ _____

20. $(a + b)^2$ _____

21. $c^2 - 2$ _____

22. $2(b + c)$ _____

       80 Activities to Make Basic Algebra Easier

ACTIVITY 6

Something of Interest

Jamal has managed to save $1,000. He keeps it hidden under his mattress. He has decided to put his money into a savings account that pays 6% interest, compounded quarterly. This means that the interest is recorded in Jamal's account every three months. At the end of the first quarter, the interest will be:

$$\$1,000 \times 6\% \times \frac{1}{4} = \$1,000 \times (6\% \div 4) = \$1,000 \times .015 = \$15.$$

> ***Remember:* Interest = Principal × Rate × Time**
> or
> **I = prt**

Directions Complete the table below to see how much Jamal has in his account at the end of one year (four quarters). Round the interest to the nearest cent.

QUARTER	P × (R × T)	INTEREST	BALANCE (PRINCIPAL + INTEREST)
1	$1,000.00 × .015	$15.00	$1,015.00 = $1,000.00 + 15.00
2	$1,015.00 × .015	$15.23	$1,030.23 = $1,015.00 + 15.23
3	$1,030.23 × .015	$15.45	$ = +
4	$	$	$ = +

Directions Complete the table below to find the total interest and total amount in a savings account that pays 8% compounded quarterly for one year. The beginning deposit is $3,000.

QUARTER	P × (R × T)	INTEREST	BALANCE (PRINCIPAL + INTEREST)
1	$3,000 × ._____		
2			
3			
4			

Total interest $ _____   Total amount $ _____

80 Activities to Make Basic Algebra Easier

ACTIVITY 7

The Compound Interest Formula

Ming has just inherited $10,000. She decides to put the money into a savings account that pays 8% interest, compounded quarterly. Ming intends to take the money out at the end of five years. How much money will be in her account at that time?

Working this problem the long way would require 20 separate problems (4 quarters × 5 years). Ming uses the compound interest formula below:

$$A = p(1 + r)^n$$

A = The total amount in savings (principal + interest).
p = The principal ($10,000 in the example).
r = The percent for *each* period (8% ÷ 4 = 2% or .02).
n = An exponent that represents the number of periods (20).

Use a scientific calculator to check Ming's work, shown below:

$$A = \$10,000 \times (1 + .02)^{20}$$
$$A = \$10,000 \times (1.02)^{20}$$
$$A = \$14,859.47 \text{ (rounded)}$$

Directions   Use a **scientific calculator** and the compound interest formula above to solve the following problems. Round your answers to the nearest cent.

1. $10,000 at 8% compounded quarterly for 4 years.

   $A = \$10,000 \times (1 + .02)^{16}$ $A = \$$ _____

2. Find the **interest** in problem one. $I = \$$ _____

3. $5,000 at 6% compounded quarterly for 5 years. ***Note:*** 6% ÷ 4 = 1.5% or .015. $A = \$$ _____

4. $15,000 at 5% compounded *semiannually* (twice a year) for 5 years. $A = \$$ _____

5. **Challenge Question:** How long would it take Ming to *double* her $10,000 compounded at 8% quarterly?

   _____ periods = _____ years.

   Another shortcut for solving compound interest problems is to use a compound interest table. Here is a *partial* table for $1.00 compounded at 2% each *period*.

6. Use the table to find the compound amount on $500 at 8% compounded quarterly for one year. _____

7. Explain how the table was made.

   _____

   _____

   _____

PERIOD	AMOUNT
1	1.02000
2	1.04040
3	1.06121
4	1.08243

12   *80 Activities to Make Basic Algebra Easier*

Activity 8

Formula Review

Working with formulas is a practical application of algebra that is related to evaluating expressions.

Example: Solve for *A* in the formula for the area of a circle if

$$A = \pi r^2, \pi \approx 3\frac{1}{7}, r = 7$$

$$A = 3\frac{1}{7} \times \frac{7}{1} \times \frac{7}{1} \qquad \text{(Substitute } 3\frac{1}{7} \text{ for } \pi \text{ and 7 for } r\text{)}$$

$$= \frac{22}{7} \times \frac{7}{1} \times \frac{7}{1} = 22 \times 7 = 154 \text{ sq. units}$$

Directions Most of the problems below involve formulas that are important in mathematics. Solve them. Then locate the answers on page 14. Connect the dots below each correct answer *in the same order in which the problems are numbered.* A drawing will appear that shows one of the places where formulas are used. (Note: Some answer dots will not be used.)

1. Perimeter of a rectangle:
   $P = 2l + 2w$
   Find *P*, if *l* = 5 and *w* = 4.
   Answer: _____

2. Area of a rectangle:
   $A = lw$
   Find *A*, if *l* = 5 and *w* = 4.
   Answer: _____

3. Area of a circle:
   $A = \pi r^2$
   Find A, if $\pi \approx 3\frac{1}{7}$ and *r* = 14.
   Answer: _____

4. Circumference of a circle:
   $C = \pi d$
   Find C, if $\pi \approx 3.14$ and *r* = 7.
   Answer: _____

5. Area of a triangle:
   $A = \frac{1}{2}bh$
   Find *A*, if *b* = 12 and *h* = 8.
   Answer: _____

6. Perimeter of a triangle:
   $P = a + b + c$
   Find *P*, if *a* = 5.2, *b* = 6.4, and *c* = 0.58.
   Answer: _____

7. Volume of a rectangular solid (box):
   $V = lwh$
   Find *V*, if *l* = 5, *w* = 3, and *h* = 6.
   Answer: _____

8. Volume of a cube:
   $V = e^3$
   Find *V*, if *e* = 4.
   Answer: _____

9. Volume of a cylinder (paper-towel tube):
   $V = \pi r^2 h$
   Find V, if $\pi \approx 3\frac{1}{7}$, *r* = 7, and *h* = 10.
   Answer: _____

10. Volume of a cone (as in ice-cream!):
    $V = \frac{1}{3}\pi r^2 h$
    Find V, if $\pi \approx 3\frac{1}{7}$, *r* = 7, and *h* = 9.
    Answer: _____

11. Volume of a sphere (ball):
    $V = \frac{4}{3}\pi r^3$
    Find V, if $\pi \approx 3\frac{1}{7}$ and *r* = 21.
    Answer: _____

12. Volume of a pyramid:
    $V = \frac{1}{3}Bh$
    Find *V*, if *B* = 18 sq. units and *h* = 3.
    Answer: _____

(continued)

Formula Review *(continued)*

Name _____

Date _____

18
.

14
.   88
.

20 .   .38,808

22   38,810
.   .

17.4
.

1.74
.

616 .   . 462

96
.

43.96 .   12.18   90   .1540
.   .

440
.

21.98   12
.   .

48 .   . 64

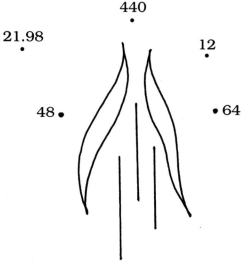

ACTIVITY 9

Basic Properties

In this activity, we will review some of the basic properties that you will use throughout algebra. (**Properties** are statements that are accepted as true without proof.) In each box below, the property is first shown using an example from arithmetic; then it is symbolized using letters.

Commutative property for addition:

If you change the *order* in which you add numbers, you still get the same sum.

$$4 + 2 = 2 + 4$$
$$a + b = b + a$$

Commutative property for multiplication:

If you change the *order* in which you multiply two numbers, you still get the same product.

$$5 \times 4 = 4 \times 5$$
$$a \times b = b \times a$$

Associative property for addition:

You can change the *grouping* of three or more numbers in addition and still get the same sum.

$$(3 + 2) + 5 = 3 + (2 + 5)$$
$$(a + b) + c = a + (b + c)$$

Associative property for multiplication:

You can change the *grouping* of three or more numbers in multiplication and still get the same product.

$$(3 \times 2) \times 6 = 3 \times (2 \times 6)$$
$$(a \times b) \times c = a \times (b \times c)$$

(continued)

Name _____

Date _____

Basic Properties *(continued)*

Identity property for addition:

When zero is added to any number, the sum is the same as the original number.

$$3 + 0 = 3$$
$$a + 0 = a$$

Identity property for multiplication:

When 1 is multiplied by any number, the product is the same as the original number.

$$5 \times 1 = 5$$
$$a \times 1 = a$$

Distributive property of multiplication with respect to addition:

In the example shown below, you can add first and then multiply, or multiply first and then add. Either way, you get the same answer.

$$3 \times (5 + 2) = 3 \times 7 = 21$$
$$OR\ (3 \times 5) + (3 \times 2) = 15 + 6 = 21$$
$$a \times (b + c) = (a \times b) + (a \times c)$$

(continued)

ACTIVITY 9

Basic Properties *(continued)*

Later on in algebra, you will use the basic properties to explain the steps that you took in solving equations. Properties can also help you to work some problems in your head. Here are some easy examples from arithmetic.

Example 1: Add 35 + 16 + 25 in your head.

  Step 1: Mentally change the order (commutative property):
35 + 25 + 16

  Step 2: 35 + 25 is easy to add in your head, so group them together (associative property):
(35 + 25) + 16 = 60 + 16 = 76

Directions 1. Show how you would use the commutative, associative, and identity properties to do this problem in your head.

  0 + 1 + 2 + 3 + 4 + 5 + 6 + 7 + 8 + 9 = _____

  Step 1: _____ (commutative)
  Step 2: _____ (associative)
  Step 3: _____ (identity)

Example 2: Work 6 × 18 in your head.

  Use the distributive property to picture the problem mentally as
$$6 \times (10 + 8) = (6 \times 10) + (6 \times 8)$$
$$= 60 + 48$$
$$= 108$$

Directions 2. Show how you would use the distributive property to do the following problems in your head:

  (a) 8 × 43 = _____

  (b) $6 \times 2\frac{1}{2}$ = _____

Directions 3. Show how you would use the distributive property with subtraction to solve:

  15 × 98 = 15 × (100 − 2) = _____

CHAPTER 2

Solving Basic Equations

This chapter will introduce students to the basic concepts involved in solving equations. Students will be asked to make and revise drawings that represent the steps in the equation-solving process. This allows them to actually see what is happening when equations are solved rather than just memorize the rules.

In order to keep students from having to think about too many things at once, all the equations in this chapter involve whole numbers. Equations involving fractions and negative numbers will be included in a later section.

Teaching Tips

Manipulative models are a wonderful way for students to experience the steps in solving equations. Unfortunately, it usually takes a lot of time to obtain and distribute the necessary equipment. Here are two suggestions to speed the process up:

1. Assemble one set of manipulatives and use an overhead projector to show your class the process. Individual students can be called up to demonstrate various problems to the class.

2. Students can simply draw equation models on paper. They can use dark circles as counters and cross them out (subtraction) or draw them in (addition) on each side of the equation (see Activities 10–18).

Overview of Activities in this Chapter

10. **Introduction to Equations**

    Overview: Introduces the concepts of the equation, using inverses to solve equations, and performing the same operation on both sides of an equation.

11. **An Equation Model**

    Materials: for each student or group, several small rectangular boxes (e.g., paper clip boxes) to represent variables; a number of

19

small counters to represent numbers; index cards on which are written the operation symbols +, −, ×, ÷, =

Overview: Uses a physical model to represent equations.

12. **Using a Model to Solve Equations (Addition and Subtraction)**

Prerequisite Activity: Activity 11, An Equation Model

Materials: for each student or group, several small rectangular boxes (e.g., paper clip boxes) to represent variables; a number of small counters to represent numbers; index cards on which are written the operation symbols +, −, ×, ÷, =

Overview: Builds on the physical model introduced in Activity 11 to demonstrate using addition and subtraction to solve equations.

13. **Solving Equations (Multiplication and Division)**

Overview: Presents the use of inverse operations to solve equations involving multiplication and division, and moves students from the physical model of Activities 11 and 12 toward a more abstract representation of algebra.

14. **Solving Equations with Whole Numbers (One Inverse Operation)**

Overview: Gives students practice solving equations with one inverse operation.

15. **Solving Equations (Several Inverse Operations)**

Overview: Building on the physical model used to represent equations, introduces the use of several inverse operations to solve equations.

16. **Solving Equations with Whole Numbers (Several Inverse Operations)**

Materials: scissors

Overview: Gives students practice solving equations using several inverse operations. As a bonus, the completed dot-to-dot puzzle forms a set of tangrams, which students are directed to form into a single large square.

17. **Solving Equations with Like Terms**

Overview: Uses a physical model of algebra to introduce the concept of combining like terms.

18. **Solving Equations (Variables on Both Sides)**

Overview: Uses the physical model of representing algebra to demonstrate solving equations with variables on both sides.

19. **Equation Review**

Overview: Uses a cross-number puzzle to give students practice using the key concepts addressed in this section.

Name _____

Date _____

Introduction to Equations

A mathematical sentence containing the = symbol is called an **equation**. Some equations look like this:

2 + 3 = 5	True
8 − 4 = 4	True
(2 × 3) + 4 = 10	True
2 + 2 = 5	False

All you can say about them is that they are true or false.

Other equations look like this:

A letter like y or x that stands for a number is called a *variable*.

$y + 2 = 5$

Read as "some number added to two equals five."

$x − 4 = 9$

Read as "some number minus four is nine."

One way to **solve equations** is to try replacements for the variable until a true sentence appears.

Example: Solve $y + 2 = 5$

Try $y = 1$: $1 + 2 = 5$ False
Try $y = 2$: $2 + 2 = 5$ False
Try $y = 3$: $3 + 2 = 5$ True

Solution: $y = 3$

Directions Solve the following equations:

1. $y + 6 = 15$ 2. $n − 4 = 15$ 3. $2y + 2 = 8$

$y =$ _____ $n =$ _____ $y =$ _____

2y means 2 times some number

I can guess the answers to those problems.

Using replacements is slow and dull. Isn't there a better way?

I thought you'd never ask! What we need is a logical approach.

(continued)

Introduction to Equations *(continued)*

Here are two basic ideas that will help you solve most equations:

1. Think of solving an equation as though you were trying to **untie a knot**. In order to untie the knot you must do the *opposite* of what the person who tied the knot did. (If they pulled the rope a certain way to form a knot, you must pull it back the *opposite* way to untie the knot.)

Study the following examples to see how you can use the idea of doing the *opposite* to solve equations.

(Mathematicians use the word *inverse* to mean opposite.)

Example 1: $n + 3 = 5$

Think: Some number has been added to 3 to equal 5. The *opposite* of adding 3 is to *subtract* 3 ($5 - 3 = 2$). Therefore, $n = 2$.

Check: Replace the variable (n) with 2 in the original equation. $2 + 3 = 5$

Example 2: $x - 4 = 9$

Think: Four has been subtracted from some number to equal 9. The *opposite* of subtracting four is to *add* 4 ($9 + 4 = 13$). Therefore, $x = 13$.

Check: $13 - 4 = 9$

(continued)

ACTIVITY 10

Introduction to Equations *(continued)*

> 2. The second idea that we will use is to think of an equation as being like the **weights on a balance scale**. The scale will balance as long as we do the same thing to *both sides* of the scale.

Example 3: 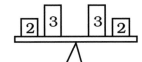 $2 + 3 = 2 + 3$

If you take the same weight *away* from each side, the scale will still balance

. . . and if you *add* the same weight to each side, the scale will still balance.

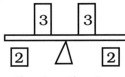 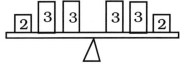

$5 - 2 = 5 - 2$ $2 + 3 + 3 = 2 + 3 + 3$

Let's put these two ideas together to show how to solve simple equations. (The logic that works for simple equations will also work for more difficult ones later on.)

Example 4:   Solve $n + 3 = 7$

Your goal is to get n equal to some number.

Since 3 is being added to n, do the opposite and subtract 3.

If we subtract 3 from *both* sides, the equation will still "balance."

$$n + 3 - 3 = 7 - 3$$
$$n + 0 = 4$$
$$n = 4$$

Check: Substitute 4 for n in the original equation. $4 + 3 = 7$

Example 5:   Solve $y - 5 = 8$

Since 5 is being subtracted, we do the opposite and *add* 5 to *both* sides of the equation.

$$y - 5 + 5 = 8 + 5$$
$$y = 13$$

Check: $13 - 5 = 8$

(continued)

ACTIVITY 10

Introduction to Equations *(continued)*

Directions   Fill in the blanks in the problems below. The first one has been done for you.

4. $n + 9 = 12$

    $n + 9 - \underline{\ 9\ } = 12 - \underline{\ 9\ }$

    $n = \underline{\ 3\ }$

    Check: $\underline{\ 3\ } + 9 = 12$

5. $x + 15 = 23$

    $x + 15 - \underline{\ \ \ \ } = 23 - \underline{\ \ \ \ }$

    $x = \underline{\ \ \ \ }$

    Check: $\underline{\ \ \ \ } + 15 = 23$

6. $y - 8 = 15$

    $y - 8 + \underline{\ \ \ \ } = 15 + \underline{\ \ \ \ }$

    $y = \underline{\ \ \ \ }$

    Check: $\underline{\ \ \ \ } - 8 = 15$

7. $a - 23 = 56$

    $a - 23 + \underline{\ \ \ \ } = 56 + \underline{\ \ \ \ }$

    $a = \underline{\ \ \ \ }$

    Check: $\underline{\ \ \ \ } - 23 = 56$

Directions   Solve the following equations. Show what must be done to *both* sides of the equation. Then check your work.

8. $n + 56 = 84$

9. $y - 93 = 213$

10. $x - 24 = 59$

11. $m + 53 = 236$

12. $a - 12 = 64$

13. $b + 64 = 235$

ACTIVITY 11

An Equation Model

Before we go any further, let's take the time to develop a physical model that will represent the equations that we want to solve. The following materials will be needed.

Materials

- Several small boxes that are the same size (empty paper clip boxes work well)
- Objects that can be counted (small stones, coins, buttons, paper clips)
- Index cards with drawings of the equal sign and various operation symbols (+, −, ×, ÷, =)

Simple equations can be represented in the following way.

Example 1: Represent $n + 3 = 5$.
The box represents the variable (n).
The objects represent the 3 and the 5.

□ + ●●● = ●●●●●

Example 2: Represent $y - 3 = 4$.

□ − ●●● = ●●●●

Example 3: Represent $2y - 3 = 5$.

□□ − ●●● = ●●●●●

Directions Make a drawing that shows how you would symbolize each of the following equations:

1. $n + 4 = 7$ 2. $2y = 6$ 3. $2x + 4 = 10$

Name _____

Date _____

Using a Model to Solve Equations (Addition and Subtraction)

The logic used to solve simple equations involving addition or subtraction can also be shown using a model.

Addition

Example 1: Solve $y + 3 = 5$.

Step 1: Symbolize $y + 3 = 5$.

$$\square + \bullet\bullet\bullet = \bullet\bullet\bullet\bullet\bullet$$

Step 2: Our goal is to get y alone on one side of the equal sign. Since 3 is being added to y, we can accomplish this by doing the *opposite* and subtracting 3 from both sides of the equation.

$$\square + \bullet\bullet\bullet = \bullet\bullet\bullet\bullet\bullet$$

> Remove (subtract) 3 objects from each side.

Answer: $\square = \bullet\bullet$ or $y = 2$

Check: $2 + 3 = 5$ ✔

Subtraction

Example 2: Solve $n - 3 = 5$.

Step 1: Symbolize $n - 3 = 5$.

$$\square - \bullet\bullet\bullet = \bullet\bullet\bullet\bullet\bullet$$

Step 2: Since 3 is being subtracted from n, do the opposite and add 3 to both sides.

$$\square = \bullet\bullet\bullet\bullet\bullet\bullet\bullet\bullet \text{ or } n = 8.$$

> Subtracting 3 from the left side and then adding it to the right side leaves n alone on the left.

Check: $8 - 3 = 5$ ✔

(continued)

ACTIVITY 12

Using a Model to Solve Equations
(Addition and Subtraction) *(continued)*

Directions   In the box below each equation, make a drawing that shows how you
would solve the equation.

1. $n + 2 = 3$

4. $y - 2 = 6$

2. $y + 4 = 6$

5. $x - 1 = 5$

3. $n + 4 = 10$

27   *80 Activities to Make Basic Algebra Easier*

ACTIVITY 13

Solving Equations (Multiplication and Division)

Multiplication

To solve an equation that involves multiplication, **do the opposite (inverse) and divide**.

Example: Solve $3y = 6$. (Read as "3 times what number is 6?")

Step 1: Divide *both* sides by 3.

$$\frac{3y}{3} = \frac{6}{3}$$

> You can show division as one numeral over another.
> ($\frac{3}{3}$ is the same as $3 \div 3$.)

$$1 \times y = 2 \qquad (3 \div 3 = 1, 6 \div 3 = 2)$$

Step 2: Write down the answer and check.

$$y = 2 \qquad (1 \times y = y)$$

Check: $3 \times 2 = 6$

The same problem can be worked using a model. For example, to symbolize $3y = 6$, we might do this:

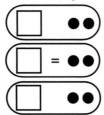

Divide *both* sides by 3. (Split both sides into 3 equal parts.)

Answer: ▢ = ●● or $y = 2$

Directions Make a drawing that shows how each of the following three problems would be solved:

1. $2y = 8$ 2. $3y = 9$ 3. $4y = 12$

(continued)

ACTIVITY 13

Solving Equations
(Multiplication and Division) *(continued)*

Division

To solve an equation that involves division, **do the opposite and multiply**.

> ***Example:*** Solve $\frac{y}{3} = 2$ (Read as "What number divided by 3 is 2?")
>
> **Step 1:** Multiply *both* sides by 3.
>
> $$\frac{\cancel{3}^{1}}{1} \times \frac{y}{\cancel{3}_{1}} = 2 \times 3 \qquad \boxed{\frac{y}{3} \text{ can be thought of as a fraction at this point.}}$$
>
> $$1 \times y = 6$$
>
> **Step 2:** Write down the answer, then check.
>
> $$y = 6 \quad \text{(Remember: } 1 \times y = y\text{)}$$
>
> ***Check:*** $\frac{6}{3} = 2$ ✔

Directions Solve the following equations. Show what you do to *both* sides of the equation.

1. $\frac{n}{4} = 3$

4. $5y = 35$

2. $\frac{c}{12} = 24$

5. $15n = 225$

3. $\frac{x}{5} = 15$

6. $\frac{b}{18} = 34$

Bonus Question:

How could you make a drawing that would show how to solve the problem below?

$$\frac{y}{3} = 2$$

Name _____

Date _____

Solving Equations with Whole Numbers (One Inverse Operation)

Directions   Solve each of the numbered math problems below. Then, use the answers to solve the cross-number puzzle.

The Inverse Brothers

Across

1. $y + 54 = 206$ $y =$ _____

3. $3x = 72$ $x =$ _____

4. $c - 12 = 418$ $c =$ _____

7. $b + 15 = 71$ $b =$ _____

8. $\frac{y}{3} = 17$ $y =$ _____

10. $n - 12 = 24$ $n =$ _____

11. $6a = 252$ $a =$ _____

12. $x^2 = 36$ $x =$ _____

Down

1. $n - 6 = 5$ $n =$ _____

2. $\frac{a}{3} = 8$ $a =$ _____

3. $4x = 824$ $x =$ _____

5. $n - 38 = 313$ $n =$ _____

6. $29x = 174$ $x =$ _____

8. $y + 38 = 570$ $y =$ _____

9. $\frac{c}{7} = 8$ $c =$ _____

10. $y - 15 = 21$ $y =$ _____

         80 Activities to Make Basic Algebra Easier

Name _____

Date _____

Solving Equations
(Several Inverse Operations)

By this time, you should be ready to solve equations that involve two or more inverse operations.

Example: Solve $2y + 3 = 9$.

Using the symbols from previous activities, we can represent the equation this way:

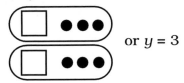

Subtract 3 from each side:

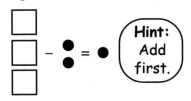

or $2y = 6$

Divide by 2:

or $y = 3$

Directions See if you can solve the following equations in the same way. Remember to show what you do to *both* sides.

1. $2y + 4 = 8$

   Hint: Subtract first.

2. $2n - 4 = 2$

   Hint: Add first.

3. $3x + 2 = 5$

   Hint: Subtract first.

4. $3y - 2 = 1$

   Hint: Add first.

5. $3y + 1 = 13$

   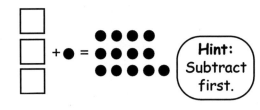

   Hint: Subtract first.

(continued)

ACTIVITY 15

Solving Equations
(Several Inverse Operations) *(continued)*

Solving equations with several inverse operations can be done more quickly if you remember this basic idea:

> When performing inverse operations with equations, do any addition or subtraction *first*, and multiplication or division next.

Example: Solve $3y - 2 = 10$

$3y - 2 + 2 = 10 + 2$ (*add* 2 to each side)

$\dfrac{3y}{3} = \dfrac{12}{3}$ (*divide* both sides by 3)

$y = 4$

Check: $(3 \times 4) - 2 = 10$

Directions Solve the following equations using the method shown in the example above. Be sure and show what you do to *both* sides of the equation.

6. $2y + 4 = 8$

7. $2n - 4 = 6$

8. $3x + 8 = 35$

9. $5n - 12 = 103$

10. $24a + 16 = 208$

11. $15c - 28 = 152$

Name _____

Date _____

Solving Equations with Whole Numbers (Several Inverse Operations)

Directions   Solve the following numbered equations. Then locate the answers below the horizontal line. If you connect the dots under each correct answer in the same order in which the problems are numbered, a drawing will appear. (Note: Some answer dots will not be used.)

Next, use scissors to cut out the 5 triangles, 1 square, and 1 parallelogram in the drawing. See if you can put the 7 pieces together to form one large square. (It may help if you go over the lines on the drawing first with a ruler.)

1. $2y + 3 = 13$ $y =$ ____
2. $4n - 2 = 30$ $n =$ ____
3. $6x + 5 = 17$ $x =$ ____
4. $2y - 12 = 6$ $y =$ ____
5. $7a - 6 = 36$ $a =$ ____

6. $3y + 8 = 44$ $y =$ ____
7. $8c - 12 = 68$ $c =$ ____
8. $3x + 6 = 15$ $x =$ ____
9. $5y - 8 = 52$ $y =$ ____
10. $15c - 12 = 3$ $c =$ ____

11. $3n + 6 = 45$ $n =$ ____
12. $8y - 15 = 73$ $y =$ ____
13. $6x + 18 = 138$ $x =$ ____
14. $b^2 - 4 = 21$ $b =$ ____

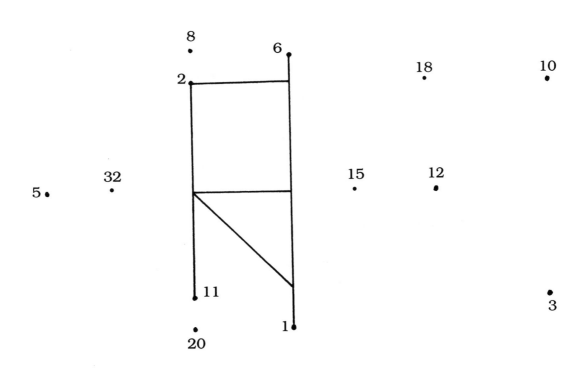

ACTIVITY 17

Solving Equations with Like Terms

Example 1: Solve $3y + 2y = 15$.

Use the symbolism from previous activities.

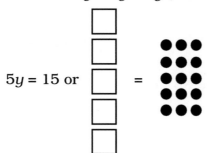

> *Like terms* have the same variables or letters and the same exponents.

$3y + 2y = 5y$ (Put all 5 squares together.)

$5y = 15$ or

> The distributive property allows you to combine like terms.
> $3y + 2y = y(3 + 2) = 5y$

Divide *both* sides by 5. (Split both sides into 5 equal parts.)

$\square = 3$ or $y = 3$

Check:
$(3 \times 3) + (2 \times 3) = 15$
$9 + 6 = 15$

Example 2: Solve $4n - 2n = 6$.

Symbolize $4n - 2n = 6$:

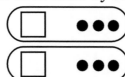

$(4n - 2n = 2n)$
$2n = 6$

Divide both sides by 2.

$\square = \bullet\bullet\bullet$ or $n = 3$ **Check:** $(4 \times 3) - (2 \times 3) = 6$
$12 - 6 = 6$ *(continued)*

ACTIVITY 17

Solving Equations with Like Terms *(continued)*

Directions Solve the following equations. Make a drawing below each of the problems using the symbols ▢ and ● .

    1. $3y - 2y = 5$ 3. $5x - 3x = 8$

    2. $2n + n = 9$ (Think of n as $1 • n$)   4. $4a - 3a = 5$

The equations in this activity are usually solved this way:

$$5y - 3y = 8$$
$$2y = 8 \quad \text{(Combine like terms.)}$$
$$\frac{2y}{2} = \frac{8}{2} \quad \text{(Divide both sides by 2.)}$$
$$y = 4 \quad \textbf{Check:} \quad (5 \times 4) - (3 \times 4) = 8$$
$$20 - 12 = 8$$

Directions Solve the following equations using the method shown in the box above. Remember to show what you do to *both* sides of the equation.

    5. $8y - 3y = 75$ 6. $9n + 3n = 144$ 7. $4c - c = 93$

         80 Activities to Make Basic Algebra Easier

Solving Equations (Variables on Both Sides)

Example: $5y = 3y + 4$

Use the symbolism from previous activities to solve equations with variables on both sides of the equals sign.

$$\square + \square + \square + \square + \square = \square + \square + \square + \begin{smallmatrix} \bullet\bullet \\ \bullet\bullet \end{smallmatrix}$$

Your goal is to get $5y$ and $3y$ together on the same side of the equation. One way to do this is to subtract $3y$ from both sides. (Cross out 3 squares on each side.)

$$\boxtimes + \boxtimes + \boxtimes + \square + \square = \boxtimes + \boxtimes + \boxtimes + \begin{smallmatrix} \bullet\bullet \\ \bullet\bullet \end{smallmatrix} \qquad (2y = 4)$$

Divide both sides by 2.

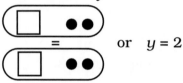

or $y = 2$

Check: $(5 \times 2) = (3 \times 2) + 4$
$$10 = 6 + 4$$

Directions Solve the following. Make a drawing below each of the problems using $\square$ or $\bullet$. Show what you do to both sides of the equation.

1. $6y = 4y + 8$ 2. $5n = 4n + 6$ 3. $5y = 2y + 6$

The equations in this activity are usually solved this way:

$$5n = 2n + 9$$
$$5n - 2n = 2n - 2n + 9$$
$$3n = 9$$
$$\frac{3n}{3} = \frac{9}{3}$$
$$n = 3 \quad \textbf{Check:} \quad (5 \times 3) = (2 \times 3) + 9$$
$$15 = 6 + 9$$

Solve the following equations using the method shown in the box above.

4. $3y = y + 12$ 5. $5n = 2n + 12$

36 *80 Activities to Make Basic Algebra Easier*

Equation Review

Directions   Solve each of the numbered math problems below. Then, use the answers to complete the cross-number puzzle.

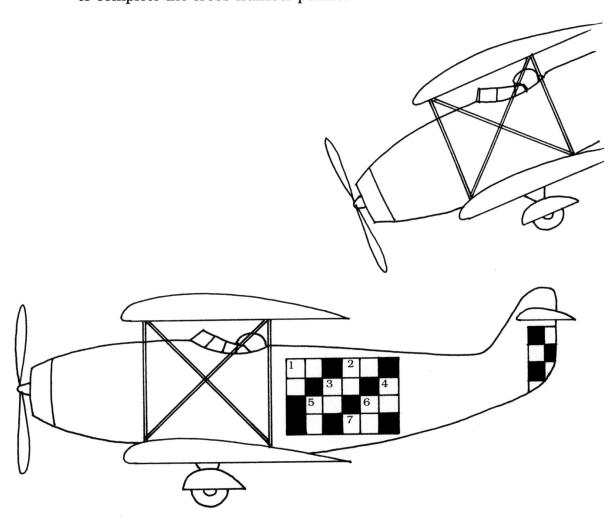

Across

1. $8y = 192$ $y =$ _____

2. $x - 8 = 4$ $x =$ _____

3. $\dfrac{a}{2} = 17$ $a =$ _____

5. $3y + 6 = 60$ $y =$ _____

6. $\dfrac{n}{15} + 26 = 29$ $n =$ _____

7. $3y = y + 38$ $y =$ _____

Down

1. $n + 28 = 54$ $n =$ _____

2. $2y + 5y = 98$ $y =$ _____

3. $2x - 23 = 53$ $x =$ _____

4. $8n = 3n + 75$ $n =$ _____

5. $4x - x = 48$ $x =$ _____

6. $n + 10 = 59$ $n =$ _____

37   *80 Activities to Make Basic Algebra Easier*

CHAPTER **3**

Working with Rational Numbers

This chapter has two objectives. First, it aims to introduce students to the concepts involved in working with positive and negative numbers through the study of integers (the set of numbers containing the elements . . . –3, –2, –1, 0, 1, 2, 3 . . .). And second, it aims to move learners toward more advanced work with equations through the study of rational numbers.

Teaching Tips

- Addition and subtraction of integers can be represented by a variety of manipulative devices. If time is a concern, a number line at the front of the room will suffice to show the process. (At the elementary level, some teachers put number lines on the floor and have students jump back and forth to represent addition and subtraction). If you have the time, it is well worth the effort to have students construct number-line rulers and solve some simple integer problems by sliding the rulers back and forth. A reproducible number-line ruler is included on page 42.

- Multiplication with integers is easy to do, but can be difficult to explain. Some students will question why a negative times a negative gives a positive result. If you feel that your students are not ready for a formal explanation using the properties, try using multiplication patterns (see Activity 28) to get this concept across.

Overview of Activities in this Chapter

20. **Introduction to Integers**

    Overview: Introduces the concepts of integers and the number line.

21. **Addition of Integers**

    Materials: for each student, two number-line rulers (page 42)

    Overview: Uses a simple number-line ruler to introduce addition of integers on the number line.

Chapter 3: Working with Rational Numbers

22. **Addition Patterns**

    Materials: for each student, two number-line rulers (page 42)

    Overview: Uses number-line rulers to guide students toward recognition of some basic patterns in addition.

23. **A Magic Circle (Addition of Integers)**

    Materials: scissors

    Overview: Uses a variant on the magic square to explore addition of integers.

24. **Addition of Integers**

    Overview: Continues the exploration of addition of integers using a variant on the magic square.

25. **Subtraction of Integers**

    Materials: for each student, two number-line rulers (page 42)

    Overview: Uses number-line rulers to introduce subtraction of integers.

26. **Subtraction Patterns**

    Materials: for each student, two number-line rulers (page 42)

    Overview: Uses number-line rulers to guide students toward recognition of some basic patterns in subtraction.

27. **Addition and Subtraction of Integers**

    Overview: Reviews addition and subtraction of integers through the use of magic squares.

28. **Multiplication Patterns (Integers)**

    Overview: Guides students toward recognition of some basic patterns in multiplication.

29. **A Division Pattern**

    Overview: Guides students toward recognition of some basic patterns in division.

30. **Working with Rational Numbers**

    Overview: Introduces the concept of rational numbers, with approaches to working with them, and a cross-number practice activity.

Teacher Guide Page

31. **Rational Number Review**

   Overview: Uses a dot-to-dot puzzle to give students practice working with rational numbers.

32. **Exponents and Integers**

   Overview: Guides students toward recognition of some patterns of exponents.

33. **An Exponent Cartoon (Zero and Negative Exponents)**

   Overview: Uses a dot-to-dot puzzle to give students practice writing equivalent expressions for numbers with negative or zero exponents.

Number-Line Rulers

You will use these number-line rulers for a number of algebra activities. Cut both rulers out carefully. Label each ruler with your name. Keep the rulers to use as you learn more about integers.

Name:	10 9 8 7 6 5 4 3 2 1 0 -1 -2 -3 -4 -5 -6 -7 -8 -9 -10

Name:	10 9 8 7 6 5 4 3 2 1 0 -1 -2 -3 -4 -5 -6 -7 -8 -9 -10

42   *80 Activities to Make Basic Algebra Easier*

Name _____

Date _____

Introduction to Integers

An integer is a member of the set
[. . . –3, –2, –1, 0, 1, 2, 3 . . .]

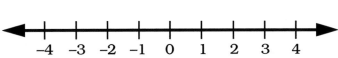

Positive integers can be symbolized
with or without the + sign.

Read –2 as "negative two," *not* as
"minus two." Read 2 or +2 as "positive
two" *not* as "plus two."

Study these examples of how to compare integers.

Examples:

5 > 3, 5 is *greater than* 3   (5 is to the *right* of 3 on the number line).

4 > –5, 4 is *greater than* –5   (4 is to the *right* of negative five).

–4 < –3, –4 is *less than* –3   (–4 is to the *left* of –3).

Directions   Use the symbols >, =, or < to complete the following statements. Above each symbol is a capital letter. Circle the correct symbol and letter for each statement. Then place the letter in the corresponding answer space at the bottom of the page. If your work is correct, a two-word message will appear. (The first one is done for you.)

1. –3 _?_ –2   N C **B**   Put "B" over (1)   6. –6 _?_ 0   P O I
   –3 < –2   > = **<**   in the answer   > = <
                                    section below.

                  A B E   T E Y
2. –6 _?_ +4   > = <   7. –3 _?_ –12   > = <

                  C A P   F O I
3. –4 _?_ 4   > = <   8. –2 _?_ +2   > = <

                  N O W   V S T
4. –3 _?_ –3   > = <   9. 0 _?_ –12   > = <

                  S E N   E B F
5. –8 _?_ –10   > = <   10. 15 _?_ –36   > = <

List the letters below:

B	__		__	__	__	__	__	__	__	__
(1)	(2)		(3)	(4)	(5)	(6)	(7)	(8)	(9)	(10)

Addition of Integers

Name _____

Date _____

For this activity you will have to make *two* identical "number-line rulers."

1. Start by taking a clean sheet of paper and creasing it the long way.

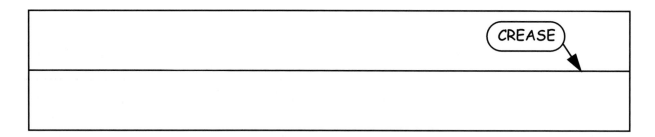

2. Use a metric ruler to mark off a number line on the crease. (Make the marks one centimeter apart.) Label the marks above *and* below the crease as shown.

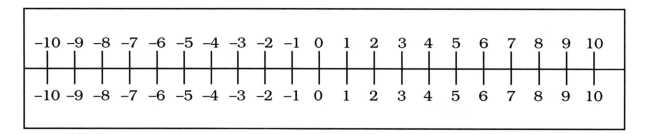

3. Cut the paper on the crease so that two "number-line rulers" are formed (see below).

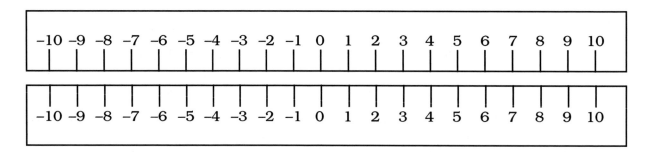

The examples on page 45 show you how to add integers with the rulers.

(continued)

ACTIVITY 21

Addition of Integers *(continued)*

For this activity, you will need two number-line rulers.

> **Example 1:** Add +3 and +2.
>
> Align the zero mark on one number line with the +3 on the other number line.
>
>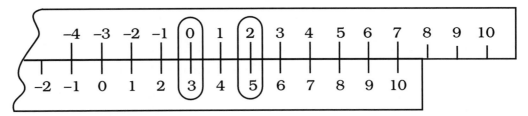
>
> Look for the answer below the +2 on the top number line: (+3) + (+2) = +5.

> **Example 2:** (–5) + (+3)
>
> Place the zero mark on one number line on top of –5 on the other number line.
>
>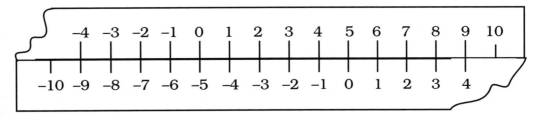
>
> Look for the answer below +3 on the top number line: (–5) + (+3) = –2.

Directions   Try these with your number-line rulers.

1. 4 + 5 _____

2. (–3) + (+6) _____

3. (–4) + (–3) _____

4. (–5) + (+5) _____

5. –4 + 6 _____

6. –6 + 4 _____

7. (–8) + (–2) _____

8. –2 + 2 _____

9. (–10) + (+3) + (+4) _____

10. –8 + ? = –3 _____

ACTIVITY 22

Addition Patterns

For this activity you will need two number-line rulers.

Directions   Add the following numbers:

1. +3 + (+2) = _____   5. –3 + (–2) = _____   9. –3 + (+2) = _____
2. +3 + (+4) = _____   6. –3 + (–4) = _____   10. +3 + (–4) = _____
3. +5 + (+4) = _____   7. –5 + (–4) = _____   11. –5 + (+4) = _____
4. +3 + (+3) = _____   8. –3 + (–3) = _____   12. –3 + (+3) = _____

Directions   Study your answers. Then complete the following:

13. In order to add integers with the *same* sign (both + or both –), you

   should _____

   _____ .

14. In order to add integers with *different* signs (one is + and one is –),

   you should _____

   _____ .

Directions   Add the following numbers:

15. –3 + (+3) = _____

16. 6 + (–6) = _____

17. –8 + (8) = _____

+2 and –2 are opposites, +3 and –3 are opposites. 0 is its own opposite.

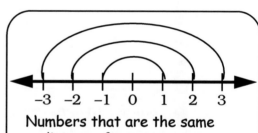

Numbers that are the same distance from zero are called *opposites*.

–*X* is read as "the opposite of *X*."

Directions   Look at your answers for problems 15–17. If you add *opposites*, the sum is

always _____ .

Name the opposite (**additive inverse**) of each of the following:

18. +12   19. –8   20. +3   21. –4   22. *a*

____   ____   ____   ____   ____

46   *80 Activities to Make Basic Algebra Easier*

Name _____

Date _____

A Magic Circle
(Addition of Integers)

Directions   Cut out the squares at the bottom of this page. Then arrange the numbers on the squares on the spaces in the puzzle so that the *sum* of the 5 squares on any line is −18.

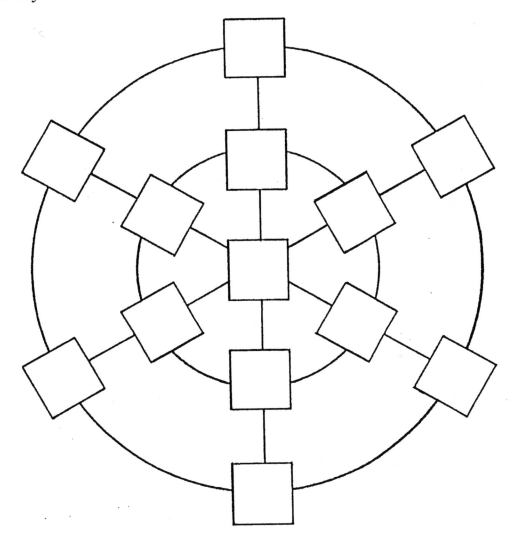

Cut out the numbered squares on the dotted lines.

| −14 | −12 | −10 | −8 | −6 | −4 | −2 | 0 | 2 | 4 | 6 | 8 | 10 |

Name _____

Date _____

Addition of Integers

Directions Place the integers –3, –2, –1, 0, 1, 2, 3, 4, 5 in the circles below so that the 3 numbers in any line add up to +3. (You may wish to write each integer on a small scrap of paper and manipulate them until you get the answer.)

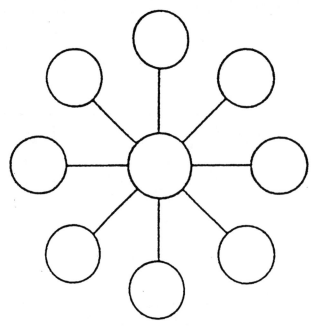

Directions Place the integers –5, –4, –3, –2, –1, 0, 1, 2, 3 in the circles below so that each side of the triangle adds up to –5.

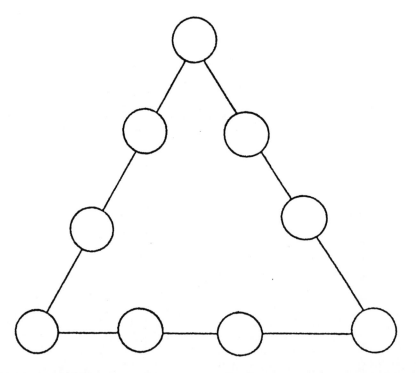

Name _____

Date _____

Subtraction of Integers

For this activity you will need two number-line rulers.

Example 1: 8 – 3 =

Place the 8 on one number line over the 3 on the other number line.

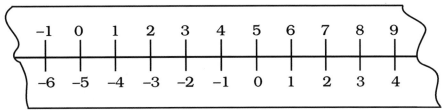

The answer is *above* the zero on the *bottom* number line: 8 – 3 = 5.

Example 2: –6 – (+4) Place the –6 over the +4.

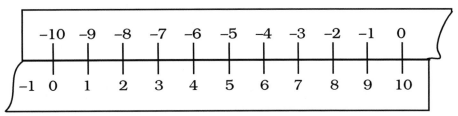

The answer is *above* the zero on the *bottom* number line: –6 – (+4) = –10.

Example 3: –6 – (–4) Place the –6 over the –4.

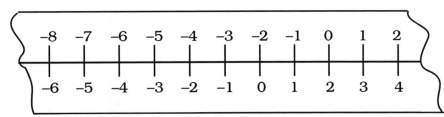

The answer is *above* the zero on the *bottom* number line: –6 – (–4) = –2.

Use your number-line rulers to solve the following:

1. 9 – 4	5. –3 – 0	9. +4 – (+9)
2. –6 – (+3)	6. 0 – (–3)	10. –5 – (+5)
3. –5 – (–3)	7. –4 – (–4)	
4. –3 – (+6)	8. –8 – (–2)	

I can check subtraction with whole numbers by addition. Would it also work for integers?

$$\begin{array}{r} 9 \\ -\ 3 \\ \hline 6 \end{array}$$

Check: 6 + 3 = 9

Try it with your answers to problems 1-10 and see.

   80 Activities to Make Basic Algebra Easier

ACTIVITY 26

Subtraction Patterns

Directions   Use your number-line rulers to solve the following subtraction and addition problems.

1. +9 − (+4)
2. −5 − (−3)
3. −6 − (+4)
4. +4 − (+9)
5. +8 − (−2)

6. +9 + (−4)
7. −5 + (+3)
8. −6 + (−4)
9. +4 + (−9)
10. +8 + (+2)

11. Study your work for problems 1–10 and look for a pattern. If you have trouble with this, review your work with opposites. In each case, how is the addition problem on the *right* related to the subtraction problem on the left?

_____

_____

To subtract with integers, *add* the *opposite* of the number that you are subtracting.

Example: −3 − (+2)

=   −3 + (−2)

=   −5

Check by addition:
if −3 − (+2) = −5
then −5 + (+2) = −3

Example: −3 − (−4)

=   −3 + (+4)

=   +1

Check: +1 + (−4) = −3

Solve the following by "adding the opposite." Check by addition.

12. +8 − (+6) = _____

13. +8 − (−6) = _____

14. −8 − (−6) = _____

15. −8 − (+6) = _____

16. −8 − 0 = _____

ACTIVITY 27

Addition and Subtraction of Integers

Directions   Fill in the missing squares so that each column, row, and diagonal adds up to the same sum.

5	−9	
	−1	
−3		−7

Sum = _____

	1	11
13	5	
	9	

Sum = _____

	−5		6
	4	3	
2	0		5
−3			−6

Sum = _____

22	−6		
		10	
8	4	2	14
	18	20	

Sum = _____

ACTIVITY 28

Multiplication Patterns (Integers)

1. Patterns can help you understand how to multiply with integers. Complete the following:

$4 \times 3 =$ _____

$4 \times 2 =$ _____

$4 \times 1 =$ _____

$4 \times 0 =$ _____

STOP

Look at the pattern formed by the answers.

Each answer is *increasing, decreasing* (circle one) by four units.

2. Use what you have learned to continue the pattern.

$4 \times (-1) =$ _____

$4 \times (-2) =$ _____

$4 \times (-3) =$ _____

0 – 4 = –4

It would seem that if we multiply a positive integer times a negative integer the answer is a *positive, negative* (circle one) integer.

3. Now try this pattern:

$4 \times (-3) =$ _____

$3 \times (-3) =$ _____

$2 \times (-3) =$ _____

$1 \times (-3) =$ _____

$0 \times (-3) =$ _____

STOP

Look at the pattern formed by the answers.

Each answer is *increasing, decreasing* (circle one) by 3 units.

(continued)

ACTIVITY 28

Multiplication Patterns
(Integers) *(continued)*

4. See if you can continue the pattern.

$$-1 \times (-3) = \text{_____}$$

$$-2 \times (-3) = \text{_____}$$

$$-3 \times (-3) = \text{_____}$$

STOP

Check your work with your teacher before going on.

Directions   Look for patterns that will help you to fill in the missing products in the multiplication table below.

×	−3	−2	−1	0	+1	+2	+3
+3				0	+3	+6	+9
+2				0	+2	+4	+6
+1				0	+1	+2	+3
0				0	0	0	0
−1							
−2							
−3							

Use the table above to help you to answer the following:

1. positive × positive is *positive/negative* (circle one)

2. positive × negative is *positive/negative* (circle one)

3. negative × positive is *positive/negative* (circle one)

4. negative × negative is *positive/negative* (circle one)

(continued)

ACTIVITY 28

Multiplication Patterns
(Integers) *(continued)*

Directions   See if you can complete this shortcut rule for multiplication of integers.

Shortcut Rule: Multiplication of Integers

In multiplication of integers, if the signs are DIFFERENT, the answer is *positive/negative* (circle one), and if the signs are the SAME, the answer is *positive/negative* (circle one).

ACTIVITY **29**

A Division Pattern

Do you remember that you can check a division problem by multiplication?

Example: $3\overline{)12}^{\,4}$ Check: $3 \times 4 = 12$

Division and multiplication are *inverse* operations.

You can use this idea to solve division problems with integers.

To solve the problem, answer the question, "What can I multiply –3 by to get +12?"

Example:

Solve: $+12 \div (-3)$

or $-3\overline{)+12}^{\,?}$

Answer: –4

Directions   Use what you have learned to solve the problems below.

1. $-4\overline{)-12}$ 2. $-4\overline{)+12}$ 3. $+4\overline{)-12}$ 4. $+4\overline{)+12}$

5. Study your answers to problems 1–4, and look for a pattern with the signs. Then complete the following by circling the word that makes each statement correct.

  (a) In division of integers if the signs are the SAME the answer is *negative, positive* (circle one).

  (b) In division of integers if the signs are DIFFERENT the answer is *positive, negative* (circle one).

These rules are similar to those for multiplication. Now use the idea of checking your work by multiplication to solve the following:

6. $-4\overline{)0}$ 7. $0\overline{)-4}$

8. Describe what happens in problem 7. _____

_____

Name _____

Date _____

Working with Rational Numbers

The same logic that is used in working with integers can be applied to rational numbers. Remember that a rational number can be expressed as the ratio of two integers. For example, $\frac{2}{3}$, $5 = \frac{5}{1}$, and $-.3 = \frac{-3}{10}$.

Example 1: $-\frac{3}{4} \cdot -\frac{2}{3} =$ [The "dot" means multiply.]

Step 1:   Multiply as you have been taught to do in arithmetic:

$$\frac{3}{4} \times \frac{2}{3} = \frac{\cancel{3}^{1}}{\cancel{4}_{2}} \times \frac{\cancel{2}^{1}}{\cancel{3}_{1}} = \frac{1}{2}$$

Step 2:   Put the correct sign on your answer (negative × negative is positive).

Answer: $+\frac{1}{2}$

Example 2: $-1.2 \div +.06$

Step 1:   Divide as you have been taught to do in arithmetic:

$$.06 \overline{)1.20} \quad \overset{20.}{}$$

Step 2:   Put the correct sign on your answer (negative divided by positive is negative).

Answer: -20

Example 3: $-4.2 - (-2.6)$

Step 1:   Remember, to subtract you must add the opposite. Rewrite the problem as $-4.2 + (+2.6)$.

Step 2:   Find the difference between 4.2 and 2.6.

$$\begin{array}{r} 4.2 \\ -\ 2.6 \\ \hline 1.6 \end{array}$$

Step 3:   Put the correct sign on your answer.

Answer: -1.6

(continued)

Name _____

Date _____

Working with Rational Numbers *(continued)*

Directions Enter answers in the cross-number puzzle below. Be sure to include + and − signs with your answers.

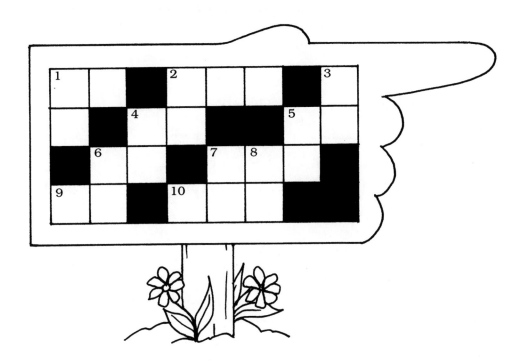

Across

1. $(-8)\,(2) =$ _____

2. $(-36)\,(-4) =$ _____

4. $(-\frac{3}{4})\,(48) =$ _____

5. $(-3)\,(-2)\,(2) =$ _____

6. $(-\frac{2}{5})\,(-100) =$ _____

7. $(-.2)^2\,(-.6) =$ _____

9. $-\frac{24}{2} =$ _____

10. $-12 \cdot [(-15) + 3] =$ _____

Down

1. $(9)\,(-2) =$ _____

2. $(-4)^2 =$ _____

3. $(-2)^5 =$ _____

4. $(150)\,(-.2) =$ _____

5. $(1400)\,(-.1)^2 =$ _____

6. $(-126) \div -3 =$ _____

7. $(-1.6) \div 40 =$ _____

8. $(-16) \div (-\frac{2}{3}) =$ _____

       80 Activities to Make Basic Algebra Easier

Name _____

Date _____

Rational Number Review

Directions   Use what you have learned about working with rational numbers to solve the following problems. Locate the answers on the dot-to-dot puzzle below. If you connect the dots below each correct answer in the same order in which the problems are numbered, a drawing will appear. (Note: Some answer dots will not be used.)

1. $-3 + (+8) =$ _____

2. $(-3) + (-5) =$ _____

3. $(-12) + 7 =$ _____

4. $(-9) + 0 =$ _____

5. $(-3) + 6 + (-3) =$ _____

6. $(-6.2) + 4.3 =$ _____

7. $(-\frac{2}{3}) + (+\frac{3}{4}) =$ _____

8. $(-9) - (-3) =$ _____

9. $(-12) - 4 =$ _____

10. $9 - (-6) =$ _____

11. $(-4) - (-11) =$ _____

12. $(-.9) - (-.23) =$ _____

13. $\frac{2}{3} - (-\frac{1}{2}) =$ _____

14. $0 - (-10) =$ _____

15. $(-3) \cdot (-4) =$ _____

16. $(-4)(5) =$ _____

17. $(-2)(-3)(-4) =$ _____

18. $(-.2)(-.3) =$ _____

19. $(-3)^2 =$ _____

20. $(-\frac{2}{3})(\frac{3}{4}) =$ _____

21. $(-12) \div 4 =$ _____

22. $(-16) \div (-4) =$ _____

23. $(-\frac{2}{3}) \div (-\frac{2}{3}) =$ _____

24. $(-\frac{2}{3}) \div \frac{3}{4} =$ _____

25. $(-\frac{1}{5}) \div (-\frac{1}{25}) =$ _____

5
•

$-\frac{8}{9}$•   1
• -5
• -8
•

•$\frac{4}{9}$ 10.5
•

-1.9
•

$-\frac{1}{2}$•
.-3 4
• -9
•   0•
•$\frac{1}{12}$

9.•

.-24 -20.   .15 -16.

.06•   24
•
-6
•

•8

.6
•

$-.14$
•

10   12   7
•   •   • .$-.67$

•$1\frac{1}{6}$

What does the cartoon have to do with positive and negative numbers?
_____

           80 Activities to Make Basic Algebra Easier

ACTIVITY 32

Exponents and Integers

Here are some patterns that will help you understand more about exponents.

1. Complete the following.

$$2^5 = 2 \times 2 \times 2 \times 2 \times 2 = \text{_____}$$
$$2^4 = 2 \times 2 \times 2 \times 2 = \text{_____}$$
$$2^3 = 2 \times 2 \times 2 = \text{_____}$$
$$2^2 = 2 \times 2 = \text{_____}$$

2. Study your work. Each answer seems to be what fraction of the previous answer? _____

3. If this pattern continues, what are the following?

$$2^1 = \text{_____}$$
$$2^0 = \text{_____}$$
$$2^{-1} = \text{_____}$$

Directions   Complete the following. Look for patterns in your answers.

Hint: Each answer is $\frac{1}{5}$ of the previous answer.

Hint: Each answer is $\frac{1}{3}$ of the previous answer.

4. $5^4 = 5 \times 5 \times 5 \times 5 = \text{\_\_\_\_}$ 5. $3^3 = 3 \times 3 \times 3 = \text{\_\_\_\_}$

$5^3 = 5 \times 5 \times 5 = \text{\_\_\_\_}$ $3^2 = 3 \times 3 = \text{\_\_\_\_}$

$5^2 = 5 \times 5 = \text{\_\_\_\_}$ $3^1 = \text{\_\_\_\_}$

$5^1 = \text{\_\_\_\_}$ $3^0 = \text{\_\_\_\_}$

$5^0 = \text{\_\_\_\_}$ $3^{-1} = \text{\_\_\_\_}$

$5^{-1} = \text{\_\_\_\_}$ $3^{-2} = \frac{1}{9} = \frac{1}{3^?}$ $? = \text{\_\_\_\_}$ ← Find the missing exponents.

$5^{-2} = \text{\_\_\_\_}$ $3^{-3} = \frac{1}{27} = \frac{1}{3^?}$ $? = \text{\_\_\_\_}$

Directions   Write using **positive** exponents.

6. $2^{-2} = \text{\_\_\_}$ 7. $2^{-3} = \text{\_\_\_}$ 8. $4^{-2} = \text{\_\_\_}$ 9. $6^{-3} = \text{\_\_\_}$ 10. $9^{-3} = \text{\_\_\_}$

11. If a is any real number ($a \neq 0$) and n is a positive integer, then

$$a^1 = \text{\_\_\_\_}$$
$$a^0 = \text{\_\_\_\_}$$
$$a^{-n} = \text{\_\_\_\_}$$

Name _____

Date _____

An Exponent Cartoon
(Zero and Negative Exponents)

Directions   Use what you have learned about exponents to solve the following problems. (Write an equivalent expression without using a zero or negative exponent.) Locate the answers below. If you connect the dots below each correct answer in the order in which the problems are numbered, a cartoon will appear. (Note: Some answer dots will not be used.)

1. $4^0 =$ _____

2. $3^{-2} =$ _____

3. $4^{-2} =$ _____

4. $-2^{-3} =$ _____

5. $(\frac{1}{4})^{-2} =$ _____

6. $3a^{-2} =$ _____

7. $4^3 \times 4^2 = 4^?$ $? =$ _____

8. $3^2 \times 3^{-3} =$ _____

9. $a^6 \div a^2 =$ _____

10. $\dfrac{a^3}{a^5} =$ _____

11. $(-15)^0 =$ _____

1
•

$\dfrac{1}{a^2}$ •

-6 • $\dfrac{1}{9}$ •   • $\dfrac{1}{6}$

$\dfrac{1}{16}$ •

$-3a^2$ •

$-\dfrac{1}{8}$ •   • 16

• 5 a^2 •

$\dfrac{3}{a^2}$ •

$\dfrac{1}{3}$ •

a^3 •

a^4 •

60   *80 Activities to Make Basic Algebra Easier*

CHAPTER 4

Equations and Inequalities with Rational Numbers

Once students have had practice in solving basic equations with whole numbers (Chapter 2) and experience with rational numbers (Chapter 3), it is time to put the two concepts together and solve equations with rational numbers.

This chapter also offers several reproducible pages with tips for solving equations with negative numbers and fractions. You may wish to distribute these to students as a handy reference for work with equations involving these elements.

Teaching Tips

This is the place in algebra where the guessing game ends. Students who have refused to show steps and give reasons for solving equations discover—to their chagrin—that they can no longer guess the answers. This is the time to do the following:

1. Start each new concept in equation solving with equations that involve only integers. The examples that follow should then include fractions and decimals.

2. Be prepared to review some arithmetic.

Overview of Activities in this Chapter

Tips on Solving Equations with Negative Numbers

An optional reproducible sheet to remind students of some of the material they have already covered.

Tips on Solving Equations with Fractions

An optional reproducible sheet to remind students of some of the material they have already covered.

Chapter 4: Equations and Inequalities with Rational Numbers

34. **Equations with Rational Numbers (One Inverse Operation)**

    Overview: Uses a cross-number puzzle to give students practice solving equations using rational numbers with one inverse operation.

35. **Equations with Rational Numbers (Several Inverse Operations)**

    Overview: Uses a magic square to give students practice solving equations using rational numbers with several inverse operations.

36. **An Equation Code (Like Terms)**

    Overview: Uses a hidden phrase and substitution code to give students practice combining like terms to solve equations.

37. **Solving Equations (Variables on Both Sides of the Equal Sign)**

    Overview: Uses a cross-number puzzle to give students practice solving equations with variables on both sides of the equal sign.

38. **Solving Inequalities**

    Overview: Introduces approaches to solving inequalities.

Tips on Solving Equations with Negative Numbers

In order to solve equations with negative numbers you will use some of these basic ideas.

Example 1: $y + (-6) = -8$

> The equation says that *adding* –6 to some number equals –8. . . . so do the opposite (inverse) and *subtract* –6 from *both* sides of the equation.

$$y + (-6) - (-6) = -8 - (-6)$$

> Remember: To subtract, you *add the opposite.*

$$y + (-6) + (+6) = -8 + (+6)$$
$$y + 0 = -8 + (+6)$$
$$y = -2$$

Check: $-2 + (-6) = -8$

Example 2: $n - (-5) = 12$

> The equation says *subtracting* –5 from some number equals 12. . . . so do the opposite and *add* –5 to *both* sides of the equation.

$$n - (-5) + (-5) = 12 + (-5)$$

> $n - (-5) = n + (+5)$

$$n + (+5) + (-5) = 12 + (-5)$$
$$n + 0 = 7$$
$$n = 7$$

Check: $7 - (-5) = 7 + (+5) = 12$

(continued)

         80 Activities to Make Basic Algebra Easier

Tips on Solving Equations with Negative Numbers *(continued)*

Example 3: $-6n = 18$

> The equation says –6 *times* some number is 18.
> . . . so do the *opposite* and *divide both* sides by –6.

$$\frac{-6n}{-6} = \frac{18}{-6}$$

$$1 \cdot n = -3$$

$$n = -3$$

Check: $-6 \cdot (-3) = 18$

Example 4: $\dfrac{n}{-3} = -12$

> The equation says some number *divided* by –3 is –12.
> . . . so do the opposite and *multiply both* sides by –3.

$$\frac{-3}{1} \cdot \frac{n}{-3} = -12 \cdot (-3)$$

$$\frac{-3}{-3} \, n = +36$$

$$1 \cdot n = 36$$

$$n = 36$$

Check: $\dfrac{36}{-3} = -12$

Tips on Solving Equations with Fractions

The same ideas that were used to solve equations with whole numbers can be used to solve equations with fractions.

Example 1: $n + \dfrac{3}{4} = \dfrac{7}{8}$

The equation says that *adding* $\dfrac{3}{4}$ to some number equals $\dfrac{7}{8}$. . . . so do the opposite and *subtract* $\dfrac{3}{4}$ from *both* sides of the equation.

$$n + \dfrac{3}{4} - \dfrac{3}{4} = \dfrac{7}{8} - \dfrac{3}{4}$$

$$n + 0 = \dfrac{7}{8} - \dfrac{6}{8}$$

$$\dfrac{3}{4} = \dfrac{6}{8}$$

$$n = \dfrac{1}{8}$$

Check: $\dfrac{1}{8} + \dfrac{3}{4} = \dfrac{1}{8} + \dfrac{6}{8} = \dfrac{7}{8}$

Example 2: $y - 2.3 = 6.2$

The equation says *subtracting* 2.3 from some number equals 6.2. . . . so do the opposite and *add* 2.3 to *both* sides.

$$y - 2.3 + 2.3 = 6.2 + 2.3$$

$$y = 8.5$$

$$\begin{array}{r} 6.2 \\ + \ 2.3 \\ \hline 8.5 \end{array}$$

Check: $8.5 - 2.3 = 6.2$

(continued)

Tips on Solving Equations with Fractions *(continued)*

Example 3: $\frac{2}{3}y = 12$

> The equation says $\frac{2}{3}$ *times* some number is 12. . . . so do the opposite and *divide both* sides by $\frac{2}{3}$.

$$\frac{3}{2} \cdot \frac{2}{3}y = \frac{12}{1} \cdot \frac{3}{2}$$

> Remember: To divide with fractions, use the *reciprocal* (multiply by $\frac{3}{2}$).

$$1 \cdot y = 18$$

$$y = 18$$

Check: $\frac{2}{3} \times \frac{18}{1} = 12$

Example 4: $.03n = 1.2$

> The equation says .03 *times* some number is 1.2. . . . so do the opposite and *divide both* sides by .03.

$$.03 \overline{)\,.03\,}^{1} \qquad \frac{.03n}{.03} = \frac{1.2}{.03} \qquad .03 \overline{)\,1.20\,}^{40.}$$

$$1 \cdot n = 40$$

$$n = 40$$

Check: $40 \times .03 = 1.20$

Name _____

Date _____

Equations with Rational Numbers (One Inverse Operation)

Directions   Solve each of the equations below. Then enter the answers in the cross-number puzzle.

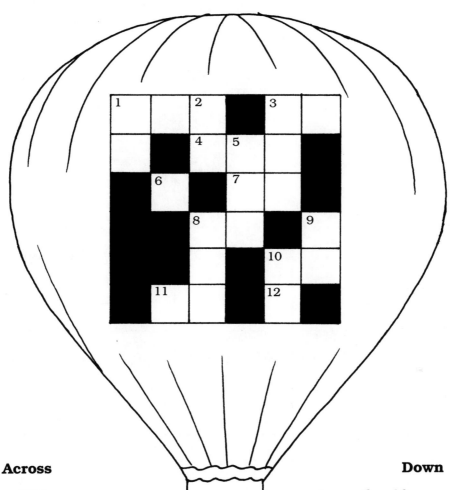

Across

1. $n + 128 = 304$ $n =$ _____

3. $\dfrac{n}{4} = 15$ $n =$ _____

4. $y + (-4) = 316$ $y =$ _____

7. $-6a = -144$ $a =$ _____

8. $y - 30 = -5$ $y =$ _____

10. $-\dfrac{3}{4}n = 33$ $n =$ _____

11. $n + 2.6 = 4$ $n =$ _____

12. $19 - n = 17$ $n =$ _____

Down

1. $y - 9 = 10$ $y =$ _____

2. $3x = 189$ $x =$ _____

3. $n + 32 = 636$ $n =$ _____

5. $x - (-5) = 230$ $x =$ _____

6. $y^2 = 36$ $y =$ _____

8. $-5n = -1120$ $n =$ _____

9. $\dfrac{2}{3}y = 16$ $y =$ _____

10. $.06a = -2.52$ $a =$ _____

67   *80 Activities to Make Basic Algebra Easier*

Name _____

Date _____

Equations with Rational Numbers (Several Inverse Operations)

Directions Use what you have learned about solving equations to work the following problems. Show what you are doing to both sides of the equation. Then write each answer on the line in the square with the problem. If you've done your work correctly, you should get the same sum every time you add the answers in a column, row, or diagonal.

> Remember to add or subtract first.

$4n + 2 = 30$ _____	$2y - 4 = 50$ _____	$3x + (-3) = 84$ _____	$5n - (-1) = 6$ _____
$\frac{1}{2}a + 1\frac{1}{2} = 10$ _____	$-2n + 30 = 4$ _____	$.03y + .67 = 1$ _____	$50 = 2t + 4$ _____
$\frac{1}{3}y - 2 = 1$ _____	$\frac{2}{3}x + 4 = 18$ _____	$2n + (-3) = 35$ _____	$\frac{3}{5}a - (-3) = 12$ _____
$3y - 4 = 89$ _____	$4y^2 + 2 = 38$ _____	$12 - 2x = 2$ _____	$\frac{n}{5} + 3 = 8$ _____

What is the sum of each column, row, and diagonal? _____

68   *80 Activities to Make Basic Algebra Easier*

Name _____

Date _____

An Equation Code (Like Terms)

A	B	C	D	E	F	G	H	I	J	K	L	M	N	O	P	Q	R	S	T	U	V	W	X	Y	Z
↓	↓	↓	↓	↓	↓	↓	↓	↓	↓	↓	↓	↓	↓	↓	↓	↓	↓	↓	↓	↓	↓	↓	↓	↓	↓
1	2	3	4	5	6	7	8	9	10	11	12	13	14	15	16	17	18	19	20	21	22	23	24	25	26

Directions Solve the following equations; show your work. Replace each answer with the appropriate letter from the table above. Then write the letter above the problem number at the bottom of the page. (The first one is done for you.)

1. $5x - 3x = 46$
$$2x = 46$$
$$\frac{2x}{2} = \frac{46}{2}$$
$$x = 23 \text{ (W)}$$

2. $2y + 3y = 75$

3. $2n + n = 54$

4. $4a - a = 33$

5. $3y - 2y + 4 = 27$

6. $6a + 3a - 2a = 63$

7. $60 = 7y - 4y$

8. $-3n - n = -32$

9. $3n + 2(n + 4) = 68$

10. $3y + (-2y) = 9$

11. $3a - 5a = -22$

12. $.02x + .03x = .25$

13. $\frac{1}{2}y + \frac{3}{4}y = 25$

14. $2(x - 3) = 4$

15. $\frac{2}{3}n - \frac{1}{3}n = 6$

16. $3(y + 2) = 45$

17. $\frac{2}{3}y + \frac{1}{3}y = 19$

18. $-2(n + 3) = -24$

19. $2b + 4(b - 5) = 94$

20. $5y - 3y - 10 = 0$

21. $.03x + 1.2x = 1.23$

22. $2(n - 20) = -2$

23. $3y - 5y = -50$

List the letters below:

W _____ _____ _____ _____   _____ _____ _____ _____   _____ _____ _____ _____
(1)   (2)   (3)   (4)   (5)   (6)   (7)   (8)   (9) (10) (11) (12)

_____ _____ _____ _____ _____   _____ _____   _____ _____ _____ _____
(13) (14) (15) (16) (17)   (18) (19)   (20) (21) (22) (23)

Name _____

Date _____

Solving Equations (Variables on Both Sides of the Equal Sign)

Just as numbers can be moved to one side of an equation by adding or subtracting, so can variables.

Directions Solve the equations below. Then use the answers to complete the cross-number puzzle.

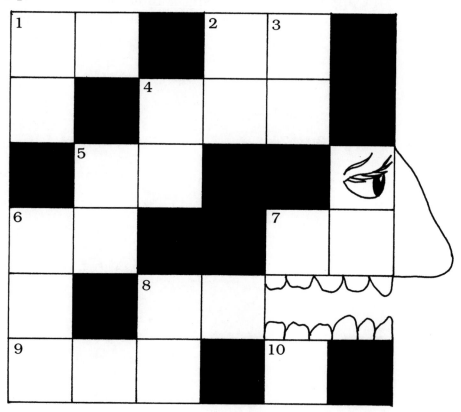

Across

1. $3x + 4 = 2x + 16$

2. $3a = 5a + 22$

4. $5a - 3a = a + 205$

5. $n - 20 = 8 - n$

6. $-3x + 4 = -2x - 19$

7. $.4y + .6y = 3y - 36$

8. $2n = .5n + 30$

9. $2(x - 6) = 3x - 218$

Down

1. $5y - 8 = y + 32$

2. $3n = 10 + 4n$

3. $4(x + 2) = 2x + 38$

4. $2y - 50 = 3y - 74$

5. $4c - 3c = 6c - 65$

6. $n - 12 = \frac{1}{2}n + 94$

8. $\frac{1}{2}x = x - 13$

10. $3a + 6 = 3a + 4$

   80 Activities to Make Basic Algebra Easier

Solving Inequalities

Name _____

Date _____

An inequality is a mathematical sentence that uses one of the following symbols:

> Read as "greater than."　　　　≥ Read as "greater than or equal to."

< Read as "less than."　　　　　≤ Read as "less than or equal to."

Can you solve inequalities just like you solve equations?

Yes, but there is *one* exception.

Example 1: $y + 3 > 5$

$$y + 3 - 3 > 5 - 3$$

$$y > 2$$

Subtract 3 from *both* sides.

There are an infinite number of numbers greater than 2. How do you check the answer?

Name the first integer greater than 2, which is 3. Put 3 in place of y and see if it makes a true statement.

Check: $y + 3 > 5$

$$3 + 3 > 5$$

$$6 > 5$$

This is too easy. What's the exception?

Try example 2.

(continued)

Solving Inequalities *(continued)*

Example 2: $-3y > 15$

$\dfrac{-3y}{-3} > \dfrac{15}{-3}$

$y > -5$

> Divide both sides by –3 or multiply by –⅓.

Check: $-3 \cdot (-4) > 15$

$12 > 15?$

> Check by using a number greater than –5. Try –4.

> What happened? It didn't work.

When you **multiply** or **divide** both sides of an inequality by a **negative** quantity you must **change** the sense (direction) of the inequality symbol. An example from arithmetic shows how this works.

Example 3: Start with the inequality $18 > 15$.

Divide *both* sides by –3:

$18 \div -3 = -6$

$15 \div -3 = -5$

$-6 < -5$ (Note that the *direction* of the inequality symbol must change to make a true sentence.)

Now, use what you have learned to solve the problem in Example 2.

Example 2, continued:

$-3y > 15$

$\dfrac{-3y}{-3} > \dfrac{15}{-3}$

$y < -5$

> Divide both sides by –3 and *reverse* the symbol.

Check: $-3 \cdot (-6) > 15$

$+18 > 15$

> Check by using a number less than –5. Try –6.

(continued)

Name _____

Date _____

Solving Inequalities *(continued)*

Directions   Solve the following equations and inequalities. Write the solution on the line next to each. Then locate each solution on the puzzle. If you connect the dots below each correct answer in the order in which the problems are numbered you will see a resident of the state in which this book is published. (Note: Some answer dots will not be used.)

1. $x + 4 = 12$ _____

2. $x + 5 > 9$ _____

3. $3y = 18$ _____

4. $5y < 25$ _____

5. $-4n = 16$ _____

6. $-3n > 21$ _____

7. $2y + 3 = 17$ _____

8. $3y + 4 < 13$ _____

9. $5a + (-6) = 4$ _____

10. $4n - (-6) = 18$ _____

11. $3y + 2y = 45$ _____

12. $4y - y > 21$ _____

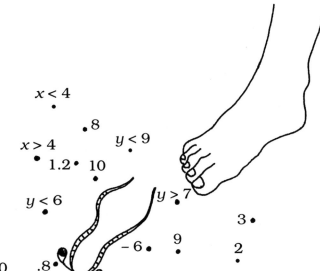
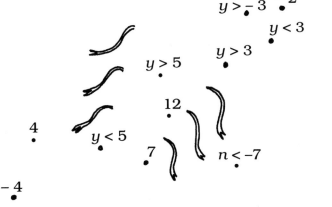

13. $3(y + 4) = -6$ _____

14. $-y + 9 = 12$ _____

15. $8y = 2y - 12$ _____

16. $5y - 8y < 9$ _____

17. $6y = 2y - 20$ _____

18. $.3y = .24$ _____

19. $-\dfrac{2}{3}y = 18$ _____

20. $n + 3 = n + 5$ _____

21. $-2y + 8 > -4$ _____

22. $12 - y = 2$ _____

23. $-y > -9$ _____

24. $n + .3 = 1.5$ _____

© 1983, 2001 J. Weston Walch, Publisher   73   *80 Activities to Make Basic Algebra Easier*

CHAPTER **5**

Polynomials

This chapter will help students understand how to add, subtract, multiply, and divide with polynomials.

Teaching Tips

This is the point where many teachers feel that they must abandon models and drawings in favor of more sophisticated explanations involving properties. We all know that students should be capable of moving to a higher level. We also know that, for some students, it will not happen—at least not yet. Why not use both approaches? Start each new concept with a drawing or model, and then move on to explanations involving the use of properties. Consider using stronger students as teachers to explain the logic of the models. Better yet, offer extra credit to students who can build models that illustrate each concept. Give students a chance to be teachers at this point, and we might all learn something new.

Overview of Activities in this Chapter

39. **Addition of Polynomials**

    Overview: Introduces addition of polynomials.

40. **Addition and Subtraction of Polynomials**

    Overview: Relates subtraction of integers to subtraction of polynomials, with a dot-to-dot puzzle for student practice.

41. **Multiplication of Monomials**

    Overview: Uses a visual model to introduce multiplication of monomials.

42. **Advanced Exponent Patterns**

    Overview: Guides students toward recognition of more complex patterns in working with exponents.

Chapter 5: Polynomials

43. **An Exponent Cross-Number Puzzle**

    Overview: Uses a cross-number puzzle to give students practice working with algebraic expressions and exponents.

44. **Multiplying a Polynomial by a Monomial**

    Overview: Introduces approaches to multiplying polynomials and monomials, including the distributive property and using visual models.

45. **A Polynomial Code (Monomial × Polynomial)**

    Overview: Uses a hidden phrase and substitution code to give students practice multiplying monomials and polynomials.

46. **Multiplying Binomials**

    Overview: Introduces methods of multiplying binomials.

47. **$(a + b)\,(a - b)$—Multiplication of Binomials (A Special Pattern)**

    Overview: Guides students toward recognition of the pattern $(a + b)\,(a - b) = a^2 - b^2$.

48. **$(a + b)^2$, $(a - b)^2$**

    Overview: Introduces approaches to squaring binomials.

49. **$(a + b)^3$—An Extra Project**

    Materials: cardboard or balsa wood, cutting tools, glue or paste

    Overview: Students build a physical model to represent $(a + b)^3$.

50. **Polynomial × Polynomial (Magic Square)**

    Overview: Uses a magic square to give students practice multiplying polynomials.

51. **Dividing a Polynomial by a Monomial**

    Overview: Introduces approaches to dividing polynomials by monomials.

52. **Dividing a Polynomial by a Binomial**

    Overview: Introduces dividing polynomials by binomials.

53. **Division of Polynomials**

    Overview: Uses a dot-to-dot puzzle to give students practice dividing polynomials.

Addition of Polynomials

> Remember, you can use the distributive property to combine *like* terms.

Example 1: $3y + 2y = y(3 + 2) = y \cdot 5 = 5y$

Example 2: $5n - 3n = n(5 - 3) = n \cdot 2 = 2n$

Example 3: $8x + (-3x) = x[8 + (-3)] = x \cdot 5 = 5x$

> Sometimes it helps to arrange *like* terms under each other.

Example 4: $3y^2 + 4y + 2 + 2y^2 - 2y - 6 =$

$$3y^2 + 4y + 2$$
$$+ \underline{2y^2 - 2y - 6}$$
$$5y^2 + 2y - 4$$

Example 5: $6a^2 + 8a - 3 + a + 9 =$

$$6a^2 + 8a - 3$$
$$+ \underline{\quad a + 9}$$
$$6a^2 + 9a + 6$$

> Think of *a* as $1 \cdot a$.

Example 6: $-3n^2 + 6n - 4 + 5n^2 - 4n - 3 =$

$$-3n^2 + 6n - 4$$
$$+ \underline{5n^2 - 4n - 3}$$
$$2n^2 + 2n - 7$$

> *Please* don't add the exponents!

(continued)

Name _____

Date _____

Addition of Polynomials *(continued)*

Directions   Solve the following problems and locate the answer below. To the left of each possible answer is a letter. If you list the letters in the same order in which the problems are numbered, the name for a very large number will appear. (NOTE: There are more letters than there are problems. Some letters will be used more than once.)

1. $(3x + 4) + (2x + 5)$ or $\begin{array}{r} 3x + 4 \\ \underline{2x + 5} \end{array}$

6. $(4x^2 + 3x + 2) + (x^2 - 5x - 6)$

2. $(2x + 7) + (x - 3)$

7. $(8x + 3y - 2z) + (-3x + 2y + z)$

3. $(8x + 1) + (-5x + 3)$

8. $(-x^2 + 5x + 4) + (6x^2 - 7x - 8)$

4. $(4x + y + 2) + (x - y + 7)$

9. $(3y^2 - 2xy) + (y^2 + 5xy)$

5. $(5x^2 + 6x - 6) + (-5x^2 - 3x + 10)$

10. $(4x^2 + 3x - 5y) + (2x - 3y)$

Possible Answers

A. $5x + 4$

B. $6x^2 + 9$

C. $3x + 10$

D. $13x + 4$

E. $4y^2 + 3xy$

F. $5x - 2y + 9$

G. $5x + 9$

H. $10x^2 + 3x + 4$

I. $4x^2 - 2x - 4$

J. $5x + 5y - 2$

K. $5x - 8y$

L. $5x^2 - 2x - 4$

M. $4x + 9$

O. $3x + 4$

P. $5x + 5y - z$

X. $4x^2 + 5x - 8y$

List the letters below:

___ ___ ___ ___ ___ ___ ___ ___ ___ ___
(1) (2) (3) (4) (5) (6) (7) (8) (9) (10)

ACTIVITY 40

Addition and Subtraction of Polynomials

In the last activity you learned how to add polynomials. Now you will learn how to subtract.

In order to understand how to subtract with polynomials, it helps to review how to subtract with integers.

Example: $-8 - (-4)$

Remember that when you subtract, you add the *opposite* of the term being subtracted. (The opposite of -4 is $+4$.)

$-8 + (+4) = -4$

Now let's use the same logic to subtract polynomials.

To subtract polynomials, add the *opposite* of each term that is being subtracted.

Example:
$6a^2 + 4b - (2a^2 + 3b)$

$$\begin{array}{r} 6a^2 + 4b \\ + \underline{-2a^2 + (-3b)} \\ 4a^2 + b \end{array}$$

Directions   Now, try the following problems to check your understanding. Show your work below the problem.

1. $5y^2 + 3 - (2y^2 - 8)$

2. $4x - (x - 2)$

3. $(x^2 + 3x - 1) - (x^2 - 2x + 1)$

4. $(5y^2 + 3y - 4) - (y^2 - 4)$

(continued)

Name _____

Date _____

Addition and Subtraction
of Polynomials *(continued)*

Add or subtract the polynomials in the problems below. If you connect the dots below each correct answer in the same order in which the problems are numbered, a cartoon will appear. (Note: Some answer dots will not be used.)

1. $(2y+2)+(3y+4)$ = _____

2. $(2y+8)+(4y-2)$ = _____

3. $(x^2+3)+(x^2-4)$ = _____

4. $(2y^2+5)+(y^2-5)$ = _____

5. $(2a+b+3)+(3a+2b+2)$ = _____

6. $(2n-m)+(-2n+m)$ = _____

7. $(3a^2+2a+4)+(a^2-a-6)$ = _____

8. $(2a^3+3a^2+2a)+(a^3-3a^2+4a+2)$ = _____

9. $(6xy+z)+(2xy-z)$ = _____

10. $(4z^3-z^2)+(z^3-3z)$ = _____

11. $(5a^3+6a^2-3a+2)+(-5a^3+a^2+3a-2)$ = _____

12. $(3y^3-2y^2)+(2y^3+y^2)+(-y^3-4y^2)$ = _____

13. $(a^3+3ab+b^2)+(2ab-b^2)$ = _____

14. $(4a+6)-(2a+4)$ = _____

15. $(3y+5x)-(y+3x)$ = _____

16. $(a-b)-(a+b)$ = _____

17. $(x^2-5x+1)-(x^2-3x+4)$ = _____

18. $(ab-xy)-(ab-3xy)$ = _____

19. $(-2y^2+3y-4)-(-2y^2+3y-2)$ = _____

20. $(3x^2+6x+10)-(3x^2+6x+6)$ = _____

21. $(4x+12)-(-4x+12)$ = _____

22. $(3a^2+6a)-(-2a^2+6a)$ = _____

23. $3x-(4x-3)$ = _____

24. $5y-(3y+2) = 10,\ \ y$ = _____

POLY NOMIAL

$2x^2 - 1$

$5a + 3b + 5$

$3y^2$

$2y + 8x$

0

$6y + 6$

$4a^2 + a - 2$

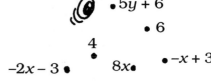

$5y + 6$

6

$2a^3 + 6a + 2$

4

$-2x - 3$

$8x$

$-x + 3$

$3a^3 + 6a + 2$

$8xy$

$2xy$

-2

$5a^2$

$a^3 + 5ab$

$-2b$

$7a^2$

$2y + 2x$

$2a + 10$

$2a + 2$

$5z^3 - z^2 - 3z$

$4y^3 - 5y^2$

       80 Activities to Make Basic Algebra Easier

Name _____

Date _____

Multiplication of Monomials

Directions   Multiplication can be represented using rectangles and cubes. Study the examples shown below. Then work the problems that follow.

Example 1:

$2 \times 3 = $ 2 ⊞ $= 6$
  3

Example 2:

$2 \times 3 \times 4 = $ [cube] $\begin{smallmatrix}2\\4\end{smallmatrix} = 24$
          3

Example 3:

$3^2 = $ 3 [grid] $= 3 \times 3 = 9$
      3

Example 4:

$x \bullet y = $ x ▭ $= xy$
          y

Example 5:

      b b
$3a \times 2b = $ [grid] $\begin{smallmatrix}a\\a\\a\end{smallmatrix} = 6ab$

Directions   What multiplication problem is represented by each of the following? (Write down the problem *and* the answer.) The first one is done for you.

1.
3 [grid] $= 3 \times 4 = 12$
   4

2.
a ▭ $= $ ____
  b

3.
4 [grid] $= $ ____
   4

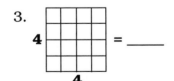

4.
y ▭ $= $ ____
   y

5. **2c** [grid] $= $ ____
       4d

6.
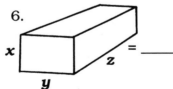
x [box] $= $ ____
     y   z

7.
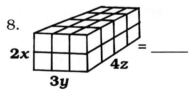
y [cube] $= $ ____
 y   y y

8.
2x [box] $= $ ____
   3y   4z

9. Why do you think that 4^2 is called "four squared" and 5^3 is called "five cubed"?

10. Make three drawings that show the difference between $3y$, $3y^2$, and $(3y)^2$.

Name _____

Date _____

Advanced Exponent Patterns

1. Complete the following: (Find the missing exponents.)

   $3^2 \times 3^3 = (3 \times 3) \times (3 \times 3 \times 3) = 3^?$ _____

   $2^1 \times 2^3 = 2 \times (2 \times 2 \times 2) = 2^?$ _____

   $5^2 \times 5^4 = (5 \times 5) \times (5 \times 5 \times 5 \times 5) = 5^?$ _____

   $y^3 \bullet y^2 = (y \bullet y \bullet y) \bullet (y \bullet y) = y^?$ _____

2. Look at the exponents in the factors above and at the exponents in each answer. It looks as if you take the exponents in the factors and _____ them to get the exponent in your answer.

3. If a is any real number and n and m are positive integers, then

   $$a^n \bullet a^m = a^? \text{ _____}$$

4. If the bases were different—for example, $3^2 \times 2^4$—would this pattern still work? _____

If you *add* exponents in multiplication with the same bases, what would you do in division?

Remember, multiplication and division are inverse (opposite) operations.

Directions   Try these:

5. $5^4 \div 5^2 = \dfrac{5^4}{5^2} = \dfrac{5 \times 5 \times 5 \times 5}{5 \times 5} = \dfrac{5}{5} \times \dfrac{5}{5} \times \dfrac{5}{1} \times \dfrac{5}{1} = 5^?$ _____

6. $3^5 \div 3^2 = \dfrac{3^5}{3^2} = \dfrac{3 \times 3 \times 3 \times 3 \times 3}{3 \times 3} = \dfrac{3}{3} \times \dfrac{3}{3} \times \dfrac{3}{1} \times \dfrac{3}{1} \times \dfrac{3}{1} = 3^?$ _____

7. From your answers above it would seem that if a is a real number and n and m are positive integers, then

   $$a^n \div a^m = a^? \text{ _____} \quad (a \neq 0)$$

8. $5^2 \div 5^2 = \dfrac{5^2}{5^2} = \dfrac{25}{25} = 1.$ $5^2 \div 5^2$ also equals 5^{2-2} or 5^0.

   Therefore $5^0 =$ _____.

9. If a is any real number ($a \neq 0$) and n is a positive integer, then

   $$a^n \div a^n = \dfrac{a^n}{a^n} = a^{n-n} = a^? = 1 \qquad ? = \text{_____} \text{ (Find the exponent.)}$$

(continued)

Name _____

Date _____

Advanced Exponent Patterns *(continued)*

Use what you have learned to solve the following:

10. $4^3 \times 4^2 = 4^?$ _____

11. $3^1 \times 3^4 = 3^?$ _____

12. $2^2 \times 2^3 \times 2^4 = 2^?$ _____

13. $8^0 =$ _____

14. $12^1 =$ _____

Study the examples below. Then work problems 15–19.

Example 1: $(3x)^2 = 3x \cdot 3x = 3 \cdot 3 \cdot x \cdot x = 9x^2$

Example 2: $(a^2b)^3 = (a^2b) \cdot (a^2b) \cdot (a^2b) = a^2 \cdot a^2 \cdot a^2 \cdot b \cdot b \cdot b = a^6b^3$

15. $(2y)^3 =$ _____ $\cdot$ _____ $\cdot$ _____ $=$ _____

16. $(3x)^3 =$ _____ $\cdot$ _____ $\cdot$ _____ $=$ _____

17. $(a^2b^3)^3 =$ _____ $\cdot$ _____ $\cdot$ _____ $=$ _____

18. $(x^2yz^3)^2 =$ _____ $\cdot$ _____ $=$ _____

19. $(c^3d)^4 =$ _____ $\cdot$ _____ $\cdot$ _____ $\cdot$ _____ $=$ _____

20. Look for an exponent pattern in problems 15–19. Then complete the following:

> If x is a positive integer, then $(ab)^x =$ _____

Study the examples below. Then work problems 21–25. Leave your answers in exponent form.

Example 3: $(5^2)^3 = 5^2 \cdot 5^2 \cdot 5^2 = 5^6$

Example 4: $(a^4)^2 = a^4 \cdot a^4 = a^8$

21. $(3^2)^4 =$ _____

22. $(y^3)^2 =$ _____

23. $(2^2)^5 =$ _____

24. $(8^2)^2 =$ _____

25. $(x^3)^3 =$ _____

26. Look for an exponent pattern in problems 21–25. Then complete the following:

> For all positive integers x and y, $(a^x)^y =$ _____

Name _____

Date _____

An Exponent Cross-Number Puzzle

$(ab)^x = a^x b^x$

$(a^x)^y = a^{xy}$

Multiply the following: (Number 4 DOWN has been done for you.)

Across

1. $(3y)^2 =$ _____

3. $(-3b)^3 =$ _____

7. $(x^3 y^4)^2 =$ _____

8. $b(bc^2 d^3)^2 =$ _____

10. $(3c^2)^2 =$ _____

12. $(-2x^2)^3 (x^2)^2 =$ _____

13. $-6y(x^2 y^2)^2 + 2x^2 y(xy^2)^2 =$

_____

Down

2. $(yx^3)^2 =$ _____

4. $7c(bc)^3 =$ _____

5. $bc^2 (bc)^2 =$ _____

6. $(y^2)^3 =$ _____

9. $(\frac{1}{2} y^2)^2 (-24y^4) =$ _____

10. $(3x^2)^2 (x^2)^3 =$ _____

11. $(x^2 y) (x^4 y^2)^2 =$ _____

80 Activities to Make Basic Algebra Easier

Multiplying a Polynomial by a Monomial

Name _____

Date _____

Example 1: Solve $3 \times (4 + 2)$

You can
ADD $3 \times 6 = 18$
FIRST

Or you can
MULTIPLY $(3 \times 4) + (3 \times 2)$
FIRST $= 12 + 6 = 18$

The distributive property says that we can do it *two* ways.

We can represent the problem using rectangles.

3 $+ 3$ $=$ $3 = 18$
4   2   4 + 2

The distributive property will help us multiply *any* polynomial by a monomial.

Example 2: $2y \, (y + 3) = 2y^2 + 6y$

$2y \, (y + 3) = 2y \boxed{2y^2} + 2y \boxed{6y} = \boxed{2y^2} + 6y \; 2y$
$\qquad\qquad\quad y \qquad\quad 3 \qquad\quad y + 3$

Directions Fill in the blanks in the problems below.

1. $3y \, (y + 4) = 3y \boxed{} + \underline{} \boxed{} = \boxed{} \,|\, \underline{} \underline{}$
    $\qquad\qquad\qquad\quad \underline{} \qquad\quad 4 \qquad\qquad \underline{} + \underline{}$

2. $4x \, (y + 6) = \underline{} \boxed{} + \underline{} \boxed{} = \boxed{} \,|\, \underline{} \underline{}$
    $\qquad\qquad\qquad\quad \underline{} \qquad\qquad \underline{} \qquad\qquad \underline{} + \underline{}$

3. $\underline{} = 3a \boxed{} + 3a \boxed{} = \boxed{} \,|\, \underline{} \underline{}$
    $\qquad\qquad\qquad\quad c \qquad\qquad 4 \qquad\qquad \underline{} + \underline{}$

4. $-5a^2 \, (a^3 + 2a) = -5a^2 \cdot \underline{} + (-5a^2) \cdot \underline{} = \underline{} + \underline{}$

Name _____

Date _____

A Polynomial Code
(Monomial × Polynomial)

Solve the following problems and locate the answers below. To the left of each possible answer is a letter. If you list the letters in the same order in which the problems are numbered, a message will appear. (NOTE: Not all letters will be used. Some letters will be used more than once.)

1. $x(x + 3)$
2. $4(x - 2)$
3. $-y(4 - 3y)$
4. $-3y^2(2y^2 - 5y)$
5. $4(2y - y^2)$
6. $2(-4 + 3)$
7. $3xy(x^2 + y^2)$
8. $y(8 - 4y)$
9. $2y + 3(y + 2)$
10. $3(2n + 1) = -9$, $n =$ _____
11. $y(-6y^3 + 15y^2)$
12. $4(x + 2)$
13. $-2y(-2x - 2y^2)$
14. $y(-4 + 3y)$
15. $(2y^2 + 3y + 4)y$
16. $3y + 2(y + 3)$
17. $3x + 3(-x + 1)$
18. $3y(-2y^3 + 5y^2)$
19. $2(2x - 4)$
20. $(2x^2 + 4x^2y + 2xy)x$
21. $5y + 2(y - 3) = 15$, $y =$ _____
22. $-x(-x - 3)$
23. $xy(3x^2 + 3y^2)$
24. $-4y + 3(2y + 4) = 8$, $y =$ _____
25. $y(2y^2 + 3y + 4)$

Possible Answers

A. $-4y + 3y^2$ H. $4x - 8$ O. $3x^3y + 3xy^3$ V. $4x - 4$
B. $x^2 + 3$ I. $2x^3 + 4x^3y + 2x^2y$ P. $8y - 4y^2$ W. $x^2 + 3x$
C. 5.6 J. 4 Q. $4xy + 4y^2$ X. $3y^2 + 6xy + 3x^2$
D. 2 K. $2y^3 + 3y^2 + 4y$ R. -2 Y. $4x + 8$
E. $5y + 6$ L. $-x^2 - 3x$ S. 3
F. $4y + 3y^2$ M. $4xy + 4y^3$ T. $-6y^4 + 15y^3$
G. -3 N. $6x + 3$ U. $x^2 - 3x$

List the letters below. Then answer the question you find hidden there.

__ __ __ __   __ __ __ __ __ __ __ __
1 2 3 4   5 6 7 8 9 10 11 12

__ __ __ __ __   __ __ __ __   __ __ __ __ ?
13 14 15 16 17   18 19 20 21   22 23 24 25

     80 Activities to Make Basic Algebra Easier

ACTIVITY 46

Multiplying Binomials

One way to understand how to multiply binomials such as $(2x + 3)(4x + 5)$ is to relate the problem to a similar one in arithmetic.

Example 1: $(20 + 3)(40 + 5)$

The fastest way is to add first, and multiply in the usual way.

$$20 + 3 = 23$$
$$40 + 5 = \times\ 45$$
$$\overline{115}$$
$$+\ 92$$
$$\overline{1035}$$

Easy, just follow the "arrows" below.

$(20 + 3)(40 + 5)$

But, suppose that we can't add first. Could we still do the problem?

$$20 \times 40 = 800$$
$$3 \times 5 = 15$$
$$20 \times 5 = 100$$
$$3 \times 40 = 120$$
$$\overline{1035}$$
} Add

Use the same logic to work this problem.

Example 2: $(2x + 3)(4x + 5)$

$(2x + 3)(4x + 5)$

$$(2x)(4x) = 8x^2$$
$$3 \times 5 = 15$$
$$(2x) \times 5 = 10x$$
$$3 \times 4x = 12x$$
} Add, combine like terms: $(10x + 12x)$

$$\overline{8x^2 + 22x + 15}$$

Here's another pattern that will work.

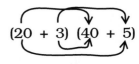

$$2x + 3$$

$$4x + 5$$

$$8x^2 + 12x$$
$$+\ 10x + 15$$
$$\overline{8x^2 + 22x + 15}$$

(continued)

87   *80 Activities to Make Basic Algebra Easier*

Multiplying Binomials *(continued)*

Directions   Solve the following. Use one of the methods you have just learned.

1. $(2 + 3)(2 + 4)$

2. $(x + 3)(x + 4)$

3. $(2x + 3)(x + 2)$

4. $(5 - 3)(5 + 2)$

5. $(y - 3)(y + 2)$

6. $(2a - 4)(3a + 2)$

7. $(6 + 2)(6 - 2)$

8. $(a + b)(a - b)$

Directions   Here is how you can multiply binomials using geometric drawings. Study the examples below. Then fill in the blanks in problems 9–10.

Example 3:

$$(20 + 3)(40 + 5) = $$

	40 + 5	
20 +	800	100
3	120	15

$$\begin{array}{r} 800 \\ 100 \\ 120 \\ +\ 15 \\ \hline 1035 \end{array}$$

 Remember that multiplication can be represented as the **area** of a **rectangular region**.

Example 4:

$$(2x + 3)(4x + 5) = $$

	4x + 5	
2x +	$8x^2$	$10x$
3	$12x$	15

$$= 8x^2 + 10x + 12x + 15 = 8x^2 + 22x + 15$$

9. $(4 + 3)(4 + 2) = $

	4 + 2	
4 +	___	___	= _____
3	___	___	= _____

10. $(x + 3)(x + 2) = $

	x + 2	
x +	___	___	= _____
3	___	___	= _____

     80 Activities to Make Basic Algebra Easier

(a + b) (a – b) — Multiplication of Binomials (A Special Pattern)

Directions   Multiply the following:

1. $(x + 2)(x - 2)$ _____

2. $(2y + 3)(2y - 3)$ _____

3. $(y + 3)(y - 3)$ _____

4. $(3a + 2)(3a - 2)$ _____

5. $(x + y)(x - y)$ _____

Study the five problems that you have worked. Then answer the following:

6. What do all the problems have in common? _____

_____

7. What pattern do you notice in the answers? _____

_____

The pattern you have observed can be symbolized this way:

$$(a + b)(a - b) = a^2 - b^2$$

This idea provides a method for the multiplication of some whole numbers in your head.

Example: $32 \times 28 = (30 + 2)(30 - 2)$

$30^2 = 900$

$2^2 = 4$

$900 - 4 = 896$

See if you can do the following in your head.

8. $(x + 3)(x - 3)$ 11. $(40 + 2)(40 - 2)$

9. $(n + 2)(n - 2)$ 12. 23×17 **Think** $(20 + 3)(20 - 3)$

10. $(2y + 4)(2y - 4)$ 13. 34×26

$(a + b)^2, (a - b)^2$

Name _____

Date _____

In order to understand how to square a binomial, it is helpful to relate the problem to a similar one in arithmetic.

Example: $(2 + 3)^2 = (2 + 3)(2 + 3)$

We know that 2 + 3 = 5, and $5^2 = 25$.

But suppose that we want to multiply first, and then add.

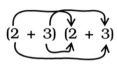

$2 \times 2 = 4$

$3 \times 3 = 9$

$2 \times 3 = 6$

$+\ 3 \times 2 = 6$ } NOTE

TOTAL = 25

The same idea can be shown using a *square*.

$(2 + 3)^2 = \begin{matrix} 2 \\ + \\ 3 \end{matrix}$

2 + 3	

$= \begin{matrix} 2 \\ + \\ 3 \end{matrix}$

2 + 3	
4	6
6	9

$= 4 + 6 + 6 + 9 = 25$

Directions   Solve the following and look for a pattern.

1. $(5 + 3)^2 = (5 + 3)(5 + 3) =$ ___25 + 15 + 15 + 9___ = _____

2. $(y + 3)^2 = (y + 3)(y + 3) =$ _____ = _____

3. $(2x + 4)^2 =$ _____ = _____ = _____

4. $(a + b)^2 =$ _____ = _____ = _____

Here's a shortcut that will help you **square binomials** in your head.

Example: $(y + 5)^2$ **Step 1:** Square the first term (y^2).

Step 2: Multiply the two terms together and double the product $(5 \cdot y = 5y, 2 \cdot 5y = 10y)$.

Step 3: Square the last term $(5^2 = 25)$.

Step 4: Add the **resulting** three terms together, i.e., $(y^2 + 10y + 25)$.

Directions   Look at your work for problems 1–4 to see if this shortcut works.

     80 Activities to Make Basic Algebra Easier

ACTIVITY 48

$(a + b)^2$, $(a - b)^2$ *(continued)*

Will the same shortcut work for $(a - b)^2$?

Try a similar problem using arithmetic and see.

Example: $(6 - 2)^2$

Try the following:

Step 1:	Square the first term:	$6^2 = 36$
Step 2:	Multiply the two terms together and double the product:	$6 \times (-2) = -12$ $(-12) \times 2 = -24$
Step 3:	Square the last term:	$(-2)^2 = +4$
Step 4:	Add the resulting three terms together:	$36 + (-24) + 4 = +16$

Check: $(6 - 2)^2 = 4^2 = 16$

Directions Solve the following problems. See if you can do them in your head using the shortcut.

5. $(a + 4)^2$ 7. $(2y + 5)^2$ 9. $(x + y)^2$

6. $(x - 3)^2$ 8. $(2y - 3)^2$ 10. $(x - y)^2$

Directions Complete the following drawings:

11.
$$5 + 2$$
$$\begin{array}{c} 5 \\ + \\ 2 \end{array}$$
= _____

12.
$$a + b$$
$$\begin{array}{c} a \\ + \\ b \end{array}$$
= _____

13. Make a drawing of a square that shows $(3 + 5)^2$.

14. Make a drawing of a square that shows $(2y + 3)^2$.

               80 Activities to Make Basic Algebra Easier

ACTIVITY 49

$(a + b)^3$ — An Extra Project

$(a + b)^3 = (a + b)\,(a + b)\,(a + b)$

First multiply $(a + b)\,(a + b)$

$(a + b)\,(a + b) = a^2 + 2ab + b^2$

Next multiply $(a^2 + 2ab + b^2)\,(a + b)$

Here's a convenient form to use:

$$a^2 + 2ab + b^2$$
$$\times \underline{\qquad\quad a + b}$$
$$a^2b + 2ab^2 + b^3$$
$$+\ \underline{a^3 + 2a^2b + ab^2\qquad\quad}$$
$$a^3 + 3a^2b + 3ab^2 + b^3$$

(Check that this is correct by using $a = 2$ and $b = 3$.)

We can represent the "cube" of a number using a cube.

Example: $3^3 =$ $= 3 \times 3 \times 3 = 27$

See if you can build a model, using a cube that represents $(a + b)^3$. It should look something like the drawing below. (**Hint:** The "cube" will consist of 8 parts.)

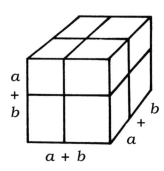

$a + b$

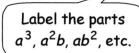

Use cardboard or balsa wood to make the 8 parts that form the "cube."

Label the parts a^3, a^2b, ab^2, etc.

Name _____

Date _____

Polynomial × Polynomial (Magic Square)

Directions Solve the following problems. The answer to each problem is in one of the squares shown below. Find the problem number that corresponds to each answer, and write it in the square. If your work is correct, the problem numbers will form a magic square.

1. $(a + 2)(a + 3) =$ _____
2. $(x + 1)(x + 2) =$ _____
3. $(y - 2)(y + 2) =$ _____
4. $(y - 2)(y - 2) =$ _____
5. $(a + 2)(a - 3) =$ _____
6. $(2x + 1)(x + 3) =$ _____
7. $(2x + 1)(x - 3) =$ _____
8. $(3y + 2)(3y + 2) =$ _____

9. $(3y + 2)(3y - 2) =$ _____
10. $(x + 1)(x - 1) =$ _____
11. $(x + y)(x + y) =$ _____
12. $(x + 3)^2 =$ _____
13. $(y + 1)(y^2 + 2y + 2) =$ _____
14. $(x^2 + y^2)(x^2 - y^2) =$ _____
15. $(a - 2)(a^2 - 2a + 1) =$ _____
16. $(x^2 - y^2)^2 =$ _____

$x^4 - 2x^2y^2 + y^4$ _____	$x^2 + 3x + 2$ _____	$y^2 - 4$ _____	$y^3 + 3y^2 + 4y + 2$ _____
$a^2 - a - 6$ _____	$x^2 + 2xy + y^2$ _____	$x^2 - 1$ _____	$9y^2 + 12y + 4$ _____
$9y^2 - 4$ _____	$2x^2 - 5x - 3$ _____	$2x^2 + 7x + 3$ _____	$x^2 + 6x + 9$ _____
$y^2 - 4y + 4$ _____	$x^4 - y^4$ _____	$a^3 - 4a^2 + 5a - 2$ _____	$a^2 + 5a + 6$ _____

What is the sum of each column, row, and diagonal? _____

© 1983, 2001 J. Weston Walch, Publisher   93   *80 Activities to Make Basic Algebra Easier*

Name _____

Date _____

Dividing a Polynomial by a Monomial

In previous activities you have seen how multiplication can be represented using rectangles.

Example 1: $2(y + 2) =$ $2\begin{array}{|c|c|}\hline y & +2 \\ \hline & \\ \hline \end{array}$ $= 2\begin{array}{|c|c|}\hline y & +2 \\ \hline 2y & 4 \\ \hline \end{array}$ $= 2y + 4$

Since division is the **inverse** (opposite) of multiplication we can use rectangles to represent dividing a polynomial by a monomial in the following way:

Example 2: $2\overline{)2y + 4}$ (This can also be written as $\dfrac{2y + 4}{2}$.)

$2\begin{array}{|c|c|}\hline \underline{} & +\underline{} \\ \hline 2y & 4 \\ \hline \end{array}$ $= 2\begin{array}{|c|c|}\hline y & +2 \\ \hline 2y & 4 \\ \hline \end{array}$ $y + 2$ is the answer.

Directions   Solve the following:

1. $3\begin{array}{|c|c|}\hline \underline{} & +\underline{} \\ \hline 6y & 9 \\ \hline \end{array}$

3. $2y\begin{array}{|c|c|}\hline \underline{} & +\underline{} \\ \hline 8y^2 & 6y \\ \hline \end{array}$

5. $4y\overline{)8y^2 + 4y} =$

2. $x\begin{array}{|c|c|}\hline \underline{} & +\underline{} \\ \hline x^2 & 3x \\ \hline \end{array}$

4. $5\begin{array}{|c|c|c|}\hline \underline{} & +\underline{} & +\underline{} \\ \hline 15y^2 & 20y & 10 \\ \hline \end{array}$

$\underline{}\begin{array}{|c|c|}\hline \underline{} & +\underline{} \\ \hline & \\ \hline \end{array} = \underline{}$

This is *too slow.* Isn't there a better way?

Sure. First try a similar problem in arithmetic.

Example: Divide first, then add.

$\dfrac{12 + 8}{4} = \dfrac{12}{4} + \dfrac{8}{4}$

$\dfrac{12}{4} + \dfrac{8}{4} = 3 + 2 = 5$

Check: Add first, then divide.
$12 + 8 = 20,\ 20 \div 4 = 5$

The distributive property makes it work.

Directions   Use what you have learned to solve the following.

6. $\dfrac{18 + 6}{3}$ 7. $\dfrac{3a + 12}{3}$ 8. $\dfrac{y^2 + 5y}{y}$ 9. $\dfrac{6n^2 + 12n}{3n}$ 10. $\dfrac{5a^3 + 10a^2 + 25a}{5a}$

(continued)

       80 Activities to Make Basic Algebra Easier

Dividing a Polynomial by a Monomial *(continued)*

Name _____

Date _____

What about a problem that has *subtraction* in it?

Try an example from arithmetic and see if it works.

Example 3: $\dfrac{12-4}{2}$ Divide first, then subtract.

$$\dfrac{12-4}{2} = \dfrac{\overset{6}{\cancel{12}}}{\underset{1}{\cancel{2}}} - \dfrac{\overset{2}{\cancel{4}}}{\underset{1}{\cancel{2}}} = 6 - 2 = 4$$

Check: Subtract first, then divide.
$12 - 4 = 8,\ 8 \div 2 = 4$

Now try a similar problem using variables.

Example 4:

$$\dfrac{4a^2-6a}{2a} = \dfrac{\overset{2a}{\cancel{4a^2}}}{\underset{1}{\cancel{2a}}} - \dfrac{\overset{3}{\cancel{6a}}}{\underset{1}{\cancel{2a}}} = 2a - 3$$

Directions Use what you have learned to solve the following:

11. $\dfrac{16n-12}{4}$

13. $\dfrac{12y^2-3y}{3y}$

15. $\dfrac{3a^2b^3-12a^2b}{3a^2b}$

12. $\dfrac{4x^2-3x}{x}$

14. $\dfrac{xy^2-x^2y}{xy}$

Sometimes the polynomial won't be *evenly* divisible by the monomial.

Example: $\dfrac{8+3}{4} = \dfrac{8}{4} + \dfrac{3}{4}$

$= 2 + \dfrac{3}{4}$

$= 2\dfrac{3}{4}$

No problem. Simply express the remainder as a fraction.

Example: $\dfrac{4x-5}{2} = \dfrac{4x}{2} - \dfrac{5}{2}$

$= 2x - \dfrac{5}{2}$

(continued)

Dividing a Polynomial by a Monomial *(continued)*

Directions   Solve the following:

16. $\dfrac{6a + 5}{3}$

18. $\dfrac{3y - 5}{y}$

17. $\dfrac{3a^2 + b}{a}$

19. $\dfrac{5 - 6y}{2}$

Here's another way to work division:

Example 5: $\dfrac{6a - 12}{2} = 2\overline{)\begin{array}{r} 3a - 6 \\ 6a - 12 \\ -6a \phantom{{}-12} \\ \hline -12 \\ -12 \\ \hline 0 \end{array}}$

Directions   Try this problem:

20. $4a\overline{)16a^2 - 12a}$

Name _____

Date _____

Dividing a Polynomial by a Binomial

In the last activity you learned how division of a **polynomial** by a **monomial** could be represented using rectangles. In this activity you will see how the problem can be worked if the divisor is a **binomial**.

Example 1: $n + 2 \overline{)n^2 + 5n + 6}$

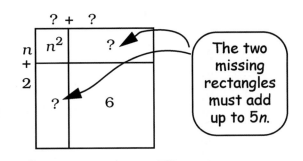

Step 1: Represent the divisor, $n + 2$, as one side of a rectangle. The other side will represent the answer.

The two missing rectangles must add up to $5n$.

Step 2: Use logic to fill in the missing terms.

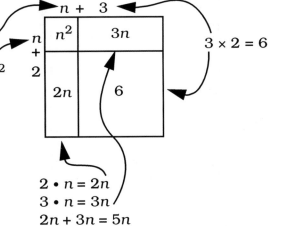

$n \times n = n^2$

$3 \times 2 = 6$

$2 \cdot n = 2n$
$3 \cdot n = 3n$
$2n + 3n = 5n$

Answer: $n + 3$

Check: $(n + 2)(n + 3) = n^2 + 5n + 6$

Directions Fill in the missing terms in the problems below.

1. $n + 3 \overline{)n^2 + 7n + 12}$

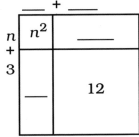

2. $n + 5 \overline{)n^2 + 7n + 10}$

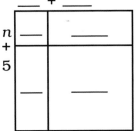

97   *80 Activities to Make Basic Algebra Easier*

ACTIVITY 52

Dividing a Polynomial
by a Binomial *(continued)*

Division with polynomials can be done in a form that is similar to division with whole numbers. Study the examples below to help you to understand how dividing with polynomials relates to arithmetic.

Example 2:

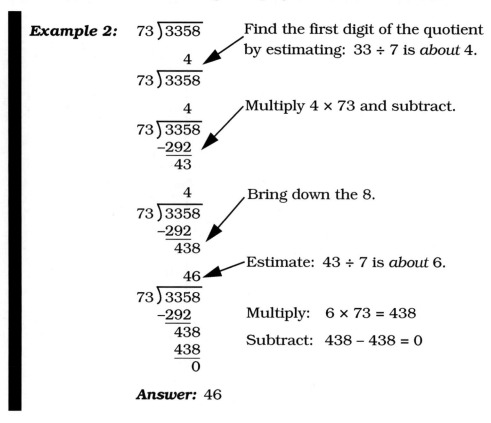

$73 \overline{)3358}$ Find the first digit of the quotient by estimating: $33 \div 7$ is *about* 4.

$$\begin{array}{r} 4 \\ 73 \overline{)3358} \end{array}$$

$$\begin{array}{r} 4 \\ 73 \overline{)3358} \\ -292 \\ \hline 43 \end{array}$$ Multiply 4×73 and subtract.

$$\begin{array}{r} 4 \\ 73 \overline{)3358} \\ -292 \\ \hline 438 \end{array}$$ Bring down the 8.

Estimate: $43 \div 7$ is *about* 6.

$$\begin{array}{r} 46 \\ 73 \overline{)3358} \\ -292 \\ \hline 438 \\ 438 \\ \hline 0 \end{array}$$

Multiply: $6 \times 73 = 438$

Subtract: $438 - 438 = 0$

Answer: 46

(continued)

ACTIVITY 52

Dividing a Polynomial
by a Binomial *(continued)*

Example 3: $n + 2 \overline{\smash{)}\, n^2 + 5n + 6}$

Estimate: $n^2 \div n = n$

$$n + 2 \overline{\smash{)}\, n^2 + 5n + 6}^{n}$$

Multiply: $n \, (n + 2)$

$$\begin{array}{r} n \\ n + 2 \overline{\smash{)}\, n^2 + 5n + 6} \\ \underline{-n^2 + 2n} \\ 3n \end{array}$$

Subtract: Remember to add the
opposite of *both* terms.

Bring down the 6. Estimate: $3n \div n = 3$

$$\begin{array}{r} n + 3 \\ n + 2 \overline{\smash{)}\, n^2 + 5n + 6} \\ \underline{-n^2 + 2n} \\ 3n + 6 \\ \underline{-3n + 6} \\ 0 \end{array}$$

Multiply: $3 \, (n + 2)$

Subtract: $3n + 6$

Answer: $n + 3$

Check by multiplication:
$(n + 3) \, (n + 2) = n^2 + 5n + 6$

Division of Polynomials

Directions Sometimes the strain of algebra is too much for some students. Solve the following problems. If you connect the dots below each correct answer in the same order in which the problems are numbered, you'll find out what happened to one student. (Note: Some answer dots will not be used.)

1. $\dfrac{4y}{y} =$ _____

2. $y^6 \div y^4 =$ _____

3. $\dfrac{y^6}{-y^3} =$ _____

4. $\dfrac{-20y^3}{-4y} =$ _____

5. $\dfrac{-12x^3}{3x} =$ _____

6. $\dfrac{4x^3y}{2x^2y} =$ _____

7. $\dfrac{4x}{12x^2} =$ _____

8. $\dfrac{(4xy)^2}{8xy^3} =$ _____

9. $\dfrac{-5x^9y^6}{15x^6y^8} =$ _____

10. $\dfrac{3x+6}{3} =$ _____

11. $\dfrac{4x-12}{4} =$ _____

12. $\dfrac{3x^2+9x}{3x} =$ _____

13. $\dfrac{4x^2+8x+12}{4} =$ _____

14. $\dfrac{2x^2+12x+18}{x+3} =$ _____

15. $\dfrac{x^2-16}{x+4} =$ _____

16. $\dfrac{4x^2}{x^2} =$ _____

$-4x^2$ •

Work the problems and see the only wild algebra student in captivity.

$2xy$ •

$2x$ •

Don't feed the student.

$\dfrac{1}{2y^2}$ •

$\dfrac{2x}{y}$ •

 • $\dfrac{1}{3x}$

• $x+2$

$5y^2$ •

$\dfrac{-x^3}{3y^2}$ •

$5y^3$ •

$x+2x+3$ •

Poor fellow, one too many tests did it.

y^3 •

• $x-3$

$-y^3$ • • $x+3$

Happens all the time.

• x^2+2x+3

y^2 • • $2x+6$

4 •

• $x-4$

80 Activities to Make Basic Algebra Easier

CHAPTER **6**

Factoring

This chapter will help students understand how to factor polynomials and how to solve equations that involve factoring.

Teaching Tips

The best way to teach students how to factor polynomials is to relate it to factoring in arithmetic. You should review the following topics in arithmetic before starting to factor polynomials:

- What are factors? Have students list sets of factors for numbers.
- Review prime factors and factor trees using examples from arithmetic.
- Define greatest common factor (GCF). Show students two ways to find the GCF (see Activity 54).
- Use arithmetic examples to review factoring and the distributive property (see Activity 55).
- Review basic percent problems before using an equation approach to find missing bases or percents.
- Teach students to constantly ask themselves, "How can I relate what I did in arithmetic to solving algebra problems?"

Overview of Activities in this Chapter

54. **Finding the GCF**

    Overview: Introduces finding the greatest common factor of monomials, with a cross-number puzzle for practice.

55. **Monomial Factors**

    Prerequisite Activity: Activity 54

    Overview: Presents using the GCF to factor polynomials, with a dot-to-dot activity for practice.

Chapter 6: Factoring

56. **A Polynomial Pattern (Factoring $x^2 + bx + c$)**

    Overview: Guides students to recognize patterns in multiplication of binomials that will help them factor polynomials in the form $x^2 + bx + c$.

57. **Factoring $x^2 + bx + c$**

    Overview: Uses a dot-to-dot puzzle to give students practice factoring polynomials in the form $x^2 + bx + c$.

58. **Factoring $ax^2 + bx + c$**

    Overview: Introduces factoring trinomials with a magic square practice exercise.

59. **An Equation Code (Solving Quadratic Equations by Factoring)**

    Overview: Introduces solving quadratic equations by factoring, with a hidden message substitution code for student practice.

ACTIVITY 54

Finding the GCF

Directions   Study the examples shown below. They will help you to recall how the greatest common factor can be found in arithmetic. (The **greatest common factor** [GCF] is the largest number that is a common factor to two or more numbers.)

> **Example 1:** Find the GCF for 8 and 20.
>
> $1 \times 8 = 8$ $1 \times 20 = 20$
> $2 \times 4 = 8$ $2 \times 10 = 20$
> The factors $4 \times 5 = 20$ Factors of 8 = {1, 2, ④, 8}
> of 8 are   The factors   Factors of 20 = {1, 2, ④, 5, 10, 20}
> 1, 2, 4, 8.   of 20 are   The GCF for 8 and 20 is 4.
>   1, 2, 4, 5, 10, 20.

> **Example 2:** Find the GCF for 2^3 and 2^5.
>
> $2^1 \times 2^2 = 2^3$ $2^1 \times 2^4 = 2^5$
> $1 \times 2^3 = 2^3$ $2^2 \times 2^3 = 2^5$ The GCF for
> The factors $1 \times 2^5 = 2^5$ 2^3 and 2^5
> of 2^3 are   The factors   is 2^3.
> $1, 2^1, 2^2, (2^3)$.   of 2^5 are
> $1, 2^1, 2^2, (2^3), 2^4, 2^5$.

Directions   Now see if you can relate what you learned in arithmetic to finding the GCF for monomials.

> **Example 3:** Find the GCF for $12x^3y^4$ and $30x^2y^6$.
>
> The GCF for 12 and 30 is 6.
> The GCF for x^3 and x^2 is x^2. The GCF for y^4 and y^6 is y^4.
> The GCF for $12x^3y^4$ and $30x^2y^6$ is $6x^2y^4$.

(continued)

Finding the GCF *(continued)*

Directions   Find the GCF for the following. Then fill in the puzzle squares with your answers. Number 7 across has been done for you.

Across

1. $36; 48$ GCF = _____

3. $39; 65$ GCF = _____

5. $4x^2; 8x$ GCF = _____

7. $5a^2b^3; 4a^3b$ GCF = _____

8. $a^4b^4c^2; a^3b^5$ GCF = _____

10. $d^3a^4c^2; d^2a^5c^3$ GCF = _____

12. $9x^2; 8y^2z^3$ GCF = _____

13. $9a^4b; 36b^2c^3$ GCF = _____

16. $c^2d; c^3d^3$ GCF = _____

18. $54c^2d^5e^3; 81d^3e^2$ GCF = _____

Down

2. $72; 48$ GCF = _____

4. $3ab; 6ab$ GCF = _____

6. $8a^3; 12a^4$ GCF = _____

7. $a^5b^4c^3; a^2b^5c^2$ GCF = _____

9. $3ab^6d^4; b^4d^2$ GCF = _____

11. $30x^2y; 24x^3$ GCF = _____

14. $5b^2c^2d^4; bc^5d^3$ GCF = _____

15. $64a^2; 160b^3$ GCF = _____

17. $6de^3; 5d^2e^2$ GCF = _____

       80 Activities to Make Basic Algebra Easier

Name _____

Date _____

Monomial Factors

In the last activity you learned how to find the greatest common factor. Now you can make use of this idea to factor polynomials.

Example 1: Write $3y^3 + 12y^2$ in factored form.
The GCF for $3y^3$ and $12y^2$ is $3y^2$.

> Once the GCF has been found, you can use the *distributive* property to write the polynomial in factored form.

$3y^2 + 12y^2 = 3y^2 (y + 4)$

Example 2: Write $16a^3 - 12a$ in factored form.
The GCF for $16a^3$ and $12a$ is $4a$.

$16a^3 - 12a =$
$4a (4a^2 - 3)$

Example 3: Write $7y^4 - 21y^3 + 14y^2$ in factored form.
The GCF for $7y^4$, $21y^3$, and $14y^2$ is $7y^2$.

$7y^4 - 21y^3 + 14y^2 =$
$7y^2 (y^2 - 3y + 2)$

Example 4: Write $4y^2 + 2y$ in factored form.
The GCF for $4y^2$ and $2y$ is $2y$.

$4y^2 + 2y =$
$2y (2y + 1)$

> Don't forget the "one"!

(continued)

105   *80 Activities to Make Basic Algebra Easier*

Name _____

Date _____

Monomial Factors (continued)

Directions Write the following in factored form. If you connect the dots below each correct answer in the same order in which the problems are numbered, a drawing will appear. (**Note:** Some answer dots will not be used. It will help if you connect the dots with a ruler.)

1. $3y^2 - 9$

2. $12y^3 + 3y$

3. $5y^3 + 15y^2$

4. $4a^2b^3 + 8ab^2$

5. $5a^2 + 5a - 15$

6. $4a^3 + 12a^2 + 16a$

7. $24x^3y^2 - 20x^2y^2 + 16xy^2$

8. $y^2(y + 2) + 3(y + 2)$

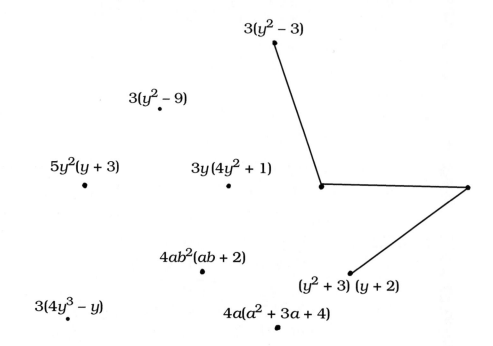

$3(y^2 - 3)$

$3(y^2 - 9)$

$5y^2(y + 3)$

$3y(4y^2 + 1)$

$4ab^2(ab + 2)$

$(y^2 + 3)(y + 2)$

$3(4y^3 - y)$

$4a(a^2 + 3a + 4)$

$5(a^2 + a - 3)$

$4xy^2(6x^2 - 5x + 4)$

Name _____

Date _____

A Polynomial Pattern
(Factoring $x^2 + bx + c$)

Directions   Use what you have learned about multiplication of binomials to complete the following table. The first problem is done for you.

$(x + r)\,(x + s)$	$x^2 + bx + c$	b	c	$r + s$	$r \times s$
$(x + 3)\,(x + 2) =$	$x^2 + 5x + 6$	5	6	$3 + 2 = 5$	$3 \times 2 = 6$
$(x + 5)\,(x + 3)$					
$(x + 4)\,(x - 2)$					
$(x - 3)\,(x - 2)$					
$(x - 3)\,(x + 2)$					

Two patterns appear in the table that will help you factor polynomials in the form $x^2 + bx + c$.

Directions   Study the table and complete the following.

    1. In each example, how does $r + s$ compare to b? _____

    2. In each example, how does $r \times s$ compare to c? _____

Example 1: Factor $x^2 + 6x + 8$

> The missing numbers must be factors of 8. 1×8 or 2×4.

> Since the middle term is $6x$, the factors of 8 must also add up to 6.

$$(x + ?)\,(x + ?)$$
$$1 + 8 = 9$$
$$2 + 4 = 6 \;\checkmark$$
Try: $(x + 2)\,(x + 4)$

Check: $(x + 2)\,(x + 4) = x^2 + 2x + 4x + 8 = x^2 + 6x + 8 \;\checkmark$

(continued)

A Polynomial Pattern
(Factoring $x^2 + bx + c$) *(continued)*

Example 2: Factor $x^2 + 3x - 10$

To factor –10, one factor must be positive and one negative. Why?

$(x + ?)(x - ?)$
The factors of –10 are:

–10, +1
+10, –1
–5, +2
+5, –2

Since the middle term is +3x, the factors of –10 must also add up to +3.

$(-10) + (+1) = -9$
$(+10) + (-1) = +9$
$(-5) + (+2) = -3$
$(+5) + (-2) = +3$ ✔

Try: $(x + 5)(x - 2)$

Check: $(x + 5)(x - 2) = x^2 + 5x - 2x - 10 = x^2 + 3x - 10$ ✔

Example 3: Factor $x^2 - 5x + 6$

The factors of +6 could be *both* positive or *both* negative. Why?

The possible factors are:

–6, –1
and
–3, –2

Add the factors:
$(-6) + (-1) = -7$
$(-3) + (-2) = -5$ ✔

Try: $(x - 3)(x - 2)$

But –5x tells us that they must both be *negative*. Why?

Check: $(x - 3)(x - 2) = x^2 - 3x - 2x + 6 = x^2 - 5x + 6$ ✔

---

         80 Activities to Make Basic Algebra Easier

ACTIVITY 57

Factoring $x^2 + bx + c$

Directions Factor the following trinomials. If you connect the dots below each correct answer in the same order in which the problems are numbered, a cartoon will appear of the "abominable math teacher." (Note: Some answer dots will not be used.)

1. $x^2 + 5x + 6$
2. $x^2 - x - 2$
3. $x^2 + 6x + 9$
4. $x^2 + x - 20$
5. $x^2 - x - 6$
6. $x^2 - 5x + 6$
7. $x^2 + 7x + 12$
8. $x^2 + 6x + 8$
9. $x^2 - 6x + 9$
10. $x^2 + 4x - 32$

11. $x^2 - 8x + 16$
12. $x^2 + 13x + 36$
13. $x^2 - 12x + 36$
14. $x^2 + 12x + 36$
15. $x^2 + 5x - 36$
16. $x^2 - 5x - 6$
17. $x^2 - 2x + 1$
18. $x^2 + 14x - 15$
19. $x^2 - 7x + 12$
20. $x^2 + 2x + 1$

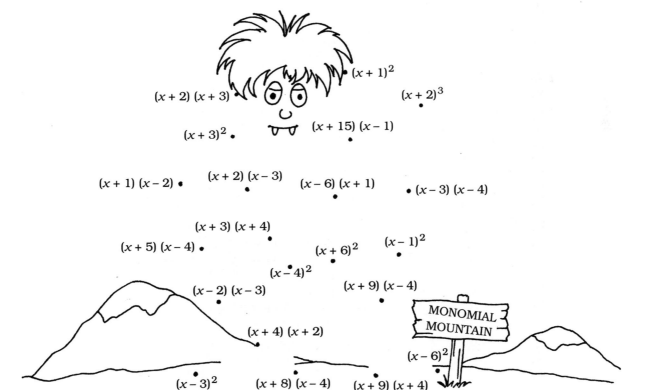

Name _____

Date _____

ACTIVITY 58

Factoring $ax^2 + bx + c$

Example 1: Factor $2x^2 + 7x + 3$

The factors of $2x^2$ must be $2x$ and x.

The factors of 3 are 3 and 1.

$(2x + ?)(x + ?)$

Try: $(2x + 3)(x + 1)$

Check: $= 2x^2 + 3x + 2x + 3$

$= 2x^2 + 5x + 3$ (doesn't check)

Try: $(2x + 1)(x + 3)$

Check: $= 2x^2 + 6x + 1x + 3$

$= 2x^2 + 7x + 3$ ✔

Example 2: Factor $4x^2 + x - 3$

The factors of $4x^2$ are $2x$ and $2x$ *or* $4x$ and x.

The factors of -3 are $+3$ and -1 *or* -3 and $+1$.

Make a chart to help your thinking.

$2x, 2x$

$4x, \ \ x$

$+3, -1$

$-3, +1$

Run through the possibilities in your head or on paper.

$(2x + 3)(2x - 1) = 4x^2 + 4x - 3$
$(2x - 3)(2x + 1) = 4x^2 - 4x - 3$
$(4x + 1)(x - 3) = 4x^2 - 11x - 3$
$(4x - 1)(x + 3) = 4x^2 + 11x - 3$
$(4x + 3)(x - 1) = 4x^2 - x - 3$
$(4x - 3)(x + 1) = 4x^2 + x - 3$ ✔

(continued)

Name _____

Date _____

Factoring $ax^2 + bx + c$ *(continued)*

Directions   Factor the following trinomials. The answer to each problem is in one of the squares shown below. Place the *problem number* in the square that corresponds to the correct answer. If your work is correct, the problem numbers will form a magic square.

1. $3y^2 + 8y + 4$

2. $4y^2 + 4y + 1$

3. $10y^2 + y - 3$

4. $4y^2 - 12y + 9$

5. $6y^2 + 13y + 6$

6. $2y^2 - y - 15$

7. $4 + 12y + 9y^2$

8. $8y^2 - 35y + 12$

9. $9y^2 - 12y + 4$

$(2y + 5)(y - 3)$ _____	$(2 + 3y)^2$ _____	$(2y + 1)(2y + 1)$ _____
$(3y + 2)(y + 2)$ _____	$(3y + 2)(2y + 3)$ _____	$(3y - 2)^2$ _____
$(8y - 3)(y - 4)$ _____	$(5y + 3)(2y - 1)$ _____	$(2y - 3)(2y - 3)$ _____

What is the sum of each column, row, and diagonal? _____

Name _____

Date _____

An Equation Code (Solving Quadratic Equations by Factoring)

One key idea helps you to understand how to solve equations by factoring. It can be represented using this equation:

$$xy = 0.$$

I know that $xy = 0$ symbolizes that two numbers multiplied together equal zero, but, who cares?

The point is that even if we don't know what x and y are, if their product is zero, either x, or y, or both *must* be zero.

This key idea can be used to solve the following problem.

Example 1: $(x - 3)(x + 2) = 0$

Either $x - 3$ or $x + 2$ must equal zero. Set each factor equal to zero.

If $x - 3 = 0$	If $x + 2 = 0$
Then $x = 3$	Then $x = -2$
Check: $3 - 3 = 0$	**Check:** $-2 + 2 = 0$

Solution: $x = 3, -2$

Now, use this idea along with what you've learned about factoring to solve this equation.

Example 2: $x^2 + x - 12 = 0$

Step 1:   Factor the left side of the equation **first**.

$(x + 4)(x - 3) = 0$

Step 2:   Set each factor equal to zero.

$x + 4 = 0 \qquad x - 3 = 0$

$x = -4 \qquad\quad x = +3$

(continued)

Name _____

Date _____

An Equation Code (Solving
Quadratic Equations by Factoring) *(continued)*

Directions   Solve the following equations. Add the solutions to each equation together. Each sum represents a letter of the alphabet (1 = A, 2 = B, 3 = C, etc.). Locate the letter that corresponds to each problem and place it in the answer space below. (The first one is done for you.) SHOW YOUR WORK!

A	B	C	D	E	F	G	H	I	J	K	L	M	N	O	P	Q	R	S	T	U	V	W	X	Y	Z
1	2	3	4	5	6	7	8	9	10	11	12	13	14	15	16	17	18	19	20	21	22	23	24	25	26

1. $x^2 - 20x + 96 = 0$
$(x - 12)(x - 8) = 0$
$x = 12, x = 8$
$12 + 8 = 20, 20 = T$

2. $x^2 - 8x + 7 = 0$

3. $x^2 - 9x + 18 = 0$

4. $x^2 - 19x - 42 = 0$

5. $x^2 - 9x - 22 = 0$

6. $x^2 - 19x + 18 = 0$

7. $x^2 - 20x + 100 = 0$

8. $x^2 - 15x + 50 = 0$

9. $x^2 - 15x - 100 = 0$

10. $x^2 - 8x + 16 = 0$

11. $x^2 - x = 0$

12. $x^2 - 18x + 81 = 0$

13. $x^2 - 4x - 12 = 0$

14. $x^2 - 6x + 9 = 0$

15. $x^2 - 15x - 16 = 0$

16. $x^2 - 18x - 40 = 0$

17. $x^2 - 13x + 42 = 0$

18. $x^2 - 5x = 0$

List the letters below:

$\underset{(1)}{\text{T}}$ $\underset{(2)}{\rule{1cm}{0.4pt}}$ $\underset{(3)}{\rule{1cm}{0.4pt}}$ $\underset{(4)}{\rule{1cm}{0.4pt}}$ $\underset{(5)}{\rule{1cm}{0.4pt}}$ $\underset{(6)}{\rule{1cm}{0.4pt}}$ $\underset{(7)}{\rule{1cm}{0.4pt}}$ $\underset{(8)}{\rule{1cm}{0.4pt}}$ $\underset{(9)}{\rule{1cm}{0.4pt}}$ $\underset{(10)}{\rule{1cm}{0.4pt}}$ $\underset{(11)}{\rule{1cm}{0.4pt}}$ $\underset{(12)}{\rule{1cm}{0.4pt}}$ $\underset{(13)}{\rule{1cm}{0.4pt}}$ $\underset{(14)}{\rule{1cm}{0.4pt}}$ $\underset{(15)}{\rule{1cm}{0.4pt}}$ $\underset{(16)}{\rule{1cm}{0.4pt}}$ $\underset{(17)}{\rule{1cm}{0.4pt}}$ $\underset{(18)}{\rule{1cm}{0.4pt}}$

80 Activities to Make Basic Algebra Easier

CHAPTER 7

Using Fractions in Algebra

This chapter will show students how arithmetic concepts concerning fractions will now apply in algebra.

Teaching Tips

The best way to introduce algebraic fractions is to review what students have previously learned about fractions in arithmetic, and connect the new material to students' prior knowledge.

- Review how to reduce fractions, then relate that concept to simplifying rational expressions (see Activity 61).

- Review basic operations with fractions (+, −, ×, ÷) and relate them to +, −, ×, ÷ with algebraic expressions.

- Review basic percent problems before using an equation approach to find missing bases or percents.

- Teach students to constantly ask themselves, "How can I relate what I did in arithmetic to solving algebra problems?"

Overview of Activities in this Chapter

60. **Rational Expressions**

   Overview: Introduces rational expressions, with a magic square practice activity.

61. **Simplifying Rational Expressions**

   Overview: Relates simplifying rational expressions to students' prior knowledge of arithmetic, with a dot-to-dot practice activity.

62. **Multiplication and Division of Fractions**

   Overview: Introduces approaches to multiplication and division of rational expressions, with a hidden message substitution code for student practice.

Chapter 7: Using Fractions in Algebra

63. **Outdoor Math (Similar Triangles)**

    Materials: Meter stick, access to outdoors area on sunny day

    Overview: Students use a meter stick to calculate the ratio of the length of the shadow of a meter stick to the stick's height, then use that ratio to determine the height of an object you assign (tree, building, flagpole, etc.).

64. **Percent Equations (Review)**

    Overview: Uses a cross-number puzzle to review percents with equations.

65. **Addition and Subtraction of Like Fractions**

    Overview: Helps students connect prior knowledge adding and subtracting fractions to adding and subtracting fractions in rational expressions, with a magic square practice activity.

66. **Addition and Subtraction of Fractions with Different Denominators**

    Overview: Builds on students' arithmetic knowledge to introduce adding and subtracting rational expressions with different denominators, with a dot-to-dot practice activity.

Rational Expressions

A **rational expression** is one that can be written in the form $\frac{a}{b}$, $(b \neq 0)$.

Examples: $\frac{3}{4}$, $\frac{5}{y}$, $\frac{a^2 + b^2}{c^2}$, $\frac{3}{1}$

What values *cannot* replace the variable in the following examples?

Example 1: $\frac{3}{n}$

Answer: $n \neq 0$ (Division by zero is undefined.)

Example 2: $\frac{6}{a - 3}$

Answer: $a \neq 3$ $(3 - 3 = 0$, the denominator cannot be zero.)

Example 3: $\frac{4y}{x^2 - 4}$

Step 1: First factor the denominator.

$$\frac{4y}{x^2 - 4} = \frac{4y}{(x + 2)\,(x - 2)}$$

Step 2: Set each factor equal to zero.

$$x + 2 = 0 \qquad x - 2 = 0$$

Solution: $x \neq -2 \qquad x \neq 2$

Directions   What values cannot replace the variable in the following problems?

1. $\frac{x}{y}$ _____   2. $\frac{5}{y + 4}$ _____   3. $\frac{3n}{x^2 + x - 6}$ _____

Check your work. Then go on to the problems on the next page. *(continued)*

       80 Activities to Make Basic Algebra Easier

Name _____

Date _____

Rational Expressions *(continued)*

Directions   Find the values that *cannot* replace the variable in each of the following expressions. If you place the numeral that stands for the problem number in the square that has the correct answer for that problem, a magic square will be formed. The first one has been done for you.

1. $\dfrac{12}{5y}$

7. $\dfrac{a}{b+8}$

13. $\dfrac{t-4}{4t-24}$

2. $\dfrac{-4}{a+2}$

8. $\dfrac{n}{4-n}$

14. $\dfrac{n}{n^2-9}$

3. $\dfrac{3}{x-5}$

9. $\dfrac{c}{9-3y}$

15. $\dfrac{a}{c(c+7)}$

4. $\dfrac{y}{3y-6}$

10. $\dfrac{y+2}{(y+4)(y-3)}$

16. $\dfrac{5}{x^2+6x+9}$

5. $\dfrac{n+2}{n-8}$

11. $\dfrac{1}{(x+2)(x+3)}$

6. $\dfrac{5}{n+4}$

12. $\dfrac{x}{x^2-x-6}$

2 ____	3, –3 ____	0, –7 ____	0 (1) ____
3 ____	–8 ____	–4 ____	–2, 3 ____
8 ____	–2, –3 ____	–4, 3 ____	4 ____
–3 ____	–2 ____	5 ____	6 ____

What is the sum of each column, row, and diagonal? Sum = _____

ACTIVITY 61

Simplifying Rational Expressions

One way to understand how to simplify rational expressions is to remember how you simplified or reduced fractions in arithmetic.

Example 1: Simplify (reduce): $\dfrac{6}{8}$

2 goes into both 6 and 8.

$$\dfrac{\overset{3}{\cancel{6}}}{\underset{4}{\cancel{8}}} = \dfrac{3}{4}$$

What you are really doing is finding the greatest common factor (GCF) for 6 and 8.

$\dfrac{6}{8} = \dfrac{3 \cdot 2}{4 \cdot 2}$

$\dfrac{2}{2}$ is another name for one.

$\dfrac{3}{4} \times 1 = \dfrac{3}{4}$

Example 2: Simplify: $\dfrac{3x^2}{15xy}$

The GCF for $3x^2$ and $15xy$ is $3x$.

$$\dfrac{\overset{1\ x}{\cancel{3x^2}}}{\underset{5\ 1}{\cancel{15xy}}} = \dfrac{x}{5y}$$

You can see why this works if you factor.

$\dfrac{3x^2}{15xy} = \dfrac{3 \cdot x \cdot x}{3 \cdot 5 \cdot x \cdot y}$

$\dfrac{3}{3}$ and $\dfrac{x}{x}$ represent one.

Example 3: Simplify: $\dfrac{x^2 + x}{x^2 + 4x + 3}$

Factor numerator *and* denominator first.

$$\dfrac{x^2 + x}{x^2 + 4x + 3} = \dfrac{x\cancel{(x+1)}}{(x+3)\cancel{(x+1)}}$$

$$= \dfrac{x}{x + 3}$$

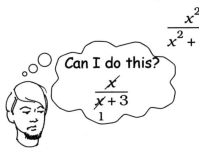

Can I do this? $\dfrac{\cancel{x}}{\underset{1}{\cancel{x}} + 3}$

No! Please don't!

(continued)

Name _____

Date _____

Simplifying Rational Expressions *(continued)*

Directions   Simplify the following expressions and locate the answers below. If you connect the dots next to each correct answer in the same order in which the problems are numbered, a cartoon will appear. (Note: Some answer dots will not be used.)

1. $\dfrac{5xy}{5x}$ _____

2. $\dfrac{x^3}{x^2 y}$ _____

3. $\dfrac{9x}{12y}$ _____

4. $\dfrac{y^6}{y^2}$ _____

5. $\dfrac{6xy^2}{18x^2 y}$ _____

6. $\dfrac{-4x^3 y}{20xy}$ _____

7. $\dfrac{y(x-y)}{(x+y)(x-y)}$ _____

8. $\dfrac{y^6(a+b)}{y^2(a+b)}$ _____

(LIFT PENCIL)

9. $\dfrac{x^2 - y^2}{x+y}$ _____

10. $\dfrac{x^2 + 5x + 6}{x^2 + x - 6}$ _____

$5x$

xy

$\dfrac{x}{y}$

$\dfrac{5x+6}{x-6}$

y

$\dfrac{x+2}{x-2}$

$-5x^2$

$\dfrac{3x}{4y}$

$3xy$

1

y^3

C

-1

$x-y$

$\dfrac{y}{3x}$

$-\dfrac{x^2}{5}$

y^4

$\dfrac{y}{x+y}$

$\dfrac{1}{x}$

$x+y$

         80 Activities to Make Basic Algebra Easier

Name _____

Date _____

Multiplication and Division of Fractions

Multiplication and division of rational expressions are easy if you can remember what you did with fractions in arithmetic.

Example 1: Multiply and simplify $\dfrac{3}{4} \times \dfrac{2}{3}$

Method A: $\dfrac{3}{4} \times \dfrac{2}{3} = \dfrac{3 \times 2}{4 \times 3} = \dfrac{\overset{1}{\cancel{6}}}{\underset{2}{\cancel{12}}} = \dfrac{1}{2}$

Shortcut method: $\dfrac{\overset{1}{\cancel{3}}}{\underset{2}{\cancel{4}}} \times \dfrac{\overset{1}{\cancel{2}}}{\underset{1}{\cancel{3}}} = \dfrac{1}{2}$

Now relate what you did in arithmetic to algebra.

Example 2: Multiply $\dfrac{6a^2}{5b} \cdot \dfrac{ab^3}{12a}$ (Write the answer in lowest terms.)

Method A: $\dfrac{6a^2}{5b} \cdot \dfrac{ab^3}{12a} = \dfrac{6a^3 b^3}{60ab} = \dfrac{a^2 b^2}{10}$

> The GCF for $6a^3 b^3$ and $60ab$ is $6ab$.

Shortcut method: $\dfrac{\overset{1a}{\cancel{6a^2}}}{\underset{1}{\cancel{5b}}} = \dfrac{a\overset{b^2}{\cancel{b^3}}}{\underset{2\ 1}{\cancel{12a}}} = \dfrac{a^2 b^2}{10}$

Example 3: Multiply $\dfrac{x^2 + 3x}{x^2 + 3x + 2} \cdot \dfrac{x^2 + 2x + 1}{x^2 + 7x + 12}$

Factor first, then use the shortcut method.

$$\dfrac{x \overset{1}{\cancel{(x+3)}}}{(x+2)\underset{1}{\cancel{(x+1)}}} \cdot \dfrac{\overset{1}{\cancel{(x+1)}}(x+1)}{\underset{1}{\cancel{(x+3)}}(x+4)} = \dfrac{x(x+1)}{(x+2)(x+4)} = \dfrac{x^2 + x}{x^2 + 6x + 8}$$

Example 4: Divide and simplify $\dfrac{3y}{x^2} \div \dfrac{6y^3}{x^2}$

> **Remember:** In division of fractions, you multiply by the *reciprocal*.
> $$\dfrac{3}{4} \div \dfrac{3}{5} = \dfrac{3}{4} \cdot \dfrac{5}{3}$$
> $$= \dfrac{5}{4} = 1\dfrac{1}{4}$$

$= \dfrac{\overset{1\ 1}{\cancel{3y}}}{\underset{1}{\cancel{x^2}}} \cdot \dfrac{\overset{1}{\cancel{x^2}}}{\underset{2\ y^2}{\cancel{6y^3}}}$

$= \dfrac{1}{2y^2}$

(continued)

Name _____

Date _____

Multiplication and Division of Fractions *(continued)*

Directions   The school's scoreboard broke down. To find out who won, solve the following problems. Locate the answer to the first problem in the table shown below. Write the letter that is above the answer in the square marked "1" in the scoreboard. Continue in the same way for problems 2–15. If your work is correct, you'll discover the winner.

A	B	N	Y	U	D	W	R	I	E	O	X	T	H
$\dfrac{x^2-4}{x+3}$	x^2	$\dfrac{x}{3}$	$\dfrac{xyz}{6}$	$\dfrac{x^2y}{3z}$	x	$\dfrac{6}{y}$	$x+3$	$\dfrac{x^2}{z^2}$	$\dfrac{x^2+xy}{y}$	$\dfrac{y^3}{4x}$	$\dfrac{x^2}{6y}$	$\dfrac{1}{x+3}$	y^2

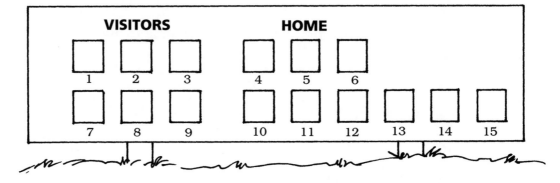

1. $\dfrac{xy}{2} \cdot \dfrac{z}{3}$ _____

2. $\dfrac{3x^2y}{8xy} \cdot \dfrac{4y^3}{6x^2}$ _____

3. $\dfrac{4x^2}{y} \cdot \dfrac{y^2}{12z}$ _____

4. $(x+2) \cdot \dfrac{(x-2)}{x+3}$ _____

5. $(x+2) \cdot \dfrac{(x+3)}{x+2}$ _____

6. $\dfrac{x+y}{x} \cdot \dfrac{x^2}{y}$ _____

7. $\dfrac{x^2-x-6}{x^2-9} \cdot \dfrac{1}{x+2}$ _____

8. $\dfrac{x^2+2x}{x} \cdot \dfrac{y^2}{x+2}$ _____

9. $\dfrac{xy}{x^2} \cdot \dfrac{x^2(x+y)}{y^2}$ _____

10. $\dfrac{x}{y} \div \dfrac{x}{6}$ _____

11. $\dfrac{xy}{z} \div \dfrac{zy}{x}$ _____

12. $\dfrac{2}{y} \div \dfrac{6}{xy}$ _____

13. $\dfrac{x^2y^2}{3} \div xy^2$ _____

14. $\dfrac{x}{y} \div \dfrac{1}{x+y}$ _____

15. $\dfrac{(x+3)(x-3)}{x-y} \div \dfrac{x-3}{x-y}$ _____

     80 Activities to Make Basic Algebra Easier

ACTIVITY 63

Outdoor Math (Similar Triangles)

You will need a meter stick for this activity. Study the example below to see how you can find the height of an object outdoors.

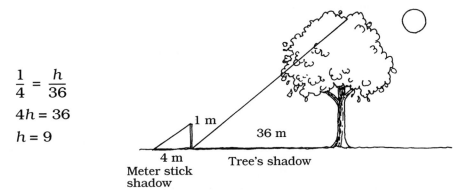

$$\frac{1}{4} = \frac{h}{36}$$

$$4h = 36$$

$$h = 9$$

1 m

36 m

4 m
Meter stick
shadow

Tree's shadow

Directions Find the height of the flagpole shown below.

Show your work here.

6 ft.

8 ft.

24 ft.

Use the meter stick to find the height of something that your teacher assigns you. Make a drawing below that shows your measurements and computation.

Name _____

Date _____

Percent Equations (Review)

Directions   Use what you have learned about equations to solve the problems shown below. Place each solution in the appropriate space in puzzle P.

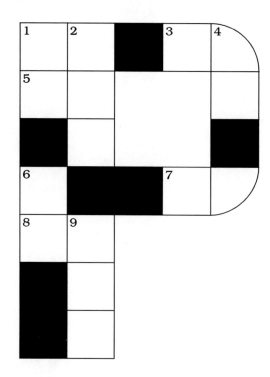

	Across		**Down**
1.	25% of 128 = _____	1.	6% of 600 = _____

1. 25% of 128 = _____

3. n% of 12 is 3 n% = _____

5. 12% of n = 7.8 n = _____

7. $66\frac{2}{3}$% of 42 = _____

8. n% of 208 = 108.16 n% = _____

Down

1. 6% of 600 = _____

2. 5% of n = 12.5 n = _____

4. n% of 18 is 9.9 n% = _____

6. $\frac{3}{5}$% of 2500 = _____

9. .4% of n = 1.024 n = _____

80 Activities to Make Basic Algebra Easier

ACTIVITY 65

Addition and Subtraction of Like Fractions

In order to add or subtract with **like fractions** (fractions with like denominators), recall what you did in arithmetic.

Addition

Example 1: Add: $\dfrac{2}{5} + \dfrac{1}{5}$

Don't add the denominators.

$$\dfrac{2}{5} + \dfrac{1}{5} = \dfrac{2+1}{5} = \dfrac{3}{5}$$

Now, use the same logic to add with rational expressions.

Example 2: Add: $\dfrac{2}{y} + \dfrac{3}{y}$

$$\dfrac{2}{y} + \dfrac{3}{y} = \dfrac{2+3}{y} = \dfrac{5}{y}$$

Subtraction   Subtraction is just as easy.

Example 3: Subtract: $\dfrac{3a}{y} - \dfrac{a}{y}$

$$\dfrac{3a}{y} - \dfrac{a}{y} = \dfrac{3a-a}{y} = \dfrac{2a}{y}$$

Example 4: Subtract: $\dfrac{2y}{3} - \dfrac{(y+6)}{3}$

$$\dfrac{2y}{3} - \dfrac{(y+6)}{3} = \dfrac{2y-(y+6)}{3}$$

$$= \dfrac{2y-y-6}{3} \longleftarrow$$ Note: *Both signs change.*

$$= \dfrac{y-6}{3}$$

Sometimes you may have to **simplify** your answer (write the answer in lowest terms).

Example 5: Add: $\dfrac{3x}{4y} + \dfrac{5x}{4y}$

$$= \dfrac{3x+5x}{4y} = \dfrac{\overset{2}{\cancel{8}}x}{\underset{1}{\cancel{4}}y} = \dfrac{2x}{y}$$

(continued)

Name _____

Date _____

Addition and Subtraction of Like Fractions *(continued)*

Directions   Solve each of the following problems. Place each answer in the square below that has the same number as the problem. (Each answer should be in lowest terms.)

1. $\dfrac{x}{2} + \dfrac{5}{2} =$ _____

2. $\dfrac{x}{2} + \dfrac{5x}{2} =$ _____

3. $\dfrac{5x+6}{y} - \dfrac{2x+7}{y} =$ _____

4. $\dfrac{y}{x} + \dfrac{5}{x} =$ _____

5. $\dfrac{5x+3}{2} - \dfrac{x+5}{2} =$ _____

6. $\dfrac{9x-4}{2} - \dfrac{5x-6}{2} =$ _____

7. $\dfrac{y+4}{3} + \dfrac{y+12}{3} =$ _____

8. $\dfrac{3x-2}{y} - \dfrac{2x-3}{y} =$ _____

9. $\dfrac{7x}{4} + \dfrac{6x}{4} - \dfrac{5x}{4} =$ _____

1	2	3
_____	_____	_____
4	**5**	**6**
_____	_____	_____
7	**8**	**9**
_____	_____	_____

Use $x = 3$ and $y = 4$ to evaluate the expressions that you have placed in the squares above. If your work is correct, the answers will form a magic square. What is the sum of each column, row, and diagonal?

Sum = _____

         80 Activities to Make Basic Algebra Easier

ACTIVITY 66

Addition and Subtraction of Fractions with Different Denominators

In order to add or subtract with fractions that have different denominators, you must first rewrite the fractions so that the denominators are the same.

Example 1: Add: $\dfrac{3}{4} + \dfrac{2}{3}$

The least common denominator (LCD) is 12.

$$\frac{3}{4} = \frac{3 \cdot 3}{3 \cdot 4} = \frac{9}{12}$$

$$\frac{2}{3} = \frac{4 \cdot 2}{4 \cdot 3} = \frac{8}{12}$$

$$\frac{9}{12} + \frac{8}{12} = \frac{17}{12} \text{ or } 1\frac{5}{12}$$

Now use the same logic to add rational expressions.

Example 2: Add: $\dfrac{3x}{2} + \dfrac{y}{3}$

The LCD is 6.

$$\frac{3x}{2} = \frac{3 \cdot 3x}{3 \cdot 2} = \frac{9x}{6}$$

$$\frac{y}{3} = \frac{2 \cdot y}{2 \cdot 3} = \frac{2y}{6}$$

$$\frac{9x}{6} + \frac{2y}{6} = \frac{9x + 2y}{6}$$

(continued)

Name _____

Date _____

Addition and Subtraction of
Fractions with Different Denominators (continued)

Example 3: Add: $\dfrac{5}{3a^3b^2} + \dfrac{3}{4ab^4}$

Both 3 and 4 will divide evenly into 12.

a will divide evenly into a^3.

b^2 will divide evenly into b^4.

The LCD for $\dfrac{5}{3a^3b^2}$ and $\dfrac{3}{4ab^4}$ is $12a^3b^4$.

$$\dfrac{5}{3a^3b^2} = \dfrac{4b^2 \cdot 5}{4b^2 \cdot 3a^3b^2} = \dfrac{20b^2}{12a^3b^4}$$

$$\dfrac{3}{4ab^4} = \dfrac{3a^2 \cdot 3}{3a^2 \cdot 4ab^4} = \dfrac{9a^2}{12a^3b^4}$$

$$\dfrac{20b^2}{12a^3b^4} + \dfrac{9a^2}{12a^3b^4} = \dfrac{20b^2 + 9a^2}{12a^3b^4}$$

Example 4: Add: $\dfrac{3x}{x^2 - y^2} + \dfrac{2y}{x + y}$

Factor the denominators first:

$x^2 - y^2 = (x + y)(x - y)$

$x + y = x + y$

$x + y$ is a factor of $x^2 - y^2$.

The LCD is $(x + y)(x - y)$.

$$\dfrac{3x}{x^2 - y^2} = \dfrac{3x}{(x + y)(x - y)}$$

$$\dfrac{2y}{x + y} = \dfrac{(x - y) \cdot 2y}{(x - y) \cdot (x + y)} = \dfrac{2xy - 2y^2}{(x - y)(x + y)}$$

$$\dfrac{3x}{(x + y)(x - y)} + \dfrac{2xy - 2y^2}{(x + y)(x - y)} = \dfrac{3x + 2xy - 2y^2}{(x + y)(x - y)}$$

$$= \dfrac{3x + 2xy - 2y^2}{x^2 - y^2}$$

(continued)

Name _____

Date _____

Addition and Subtraction of
Fractions with Different Denominators *(continued)*

Directions   This fisherman has a problem. Simplify the following problems and locate the answers below. If you connect the dots below each correct answer in the same order in which the problems are numbered, you'll see what's wrong. (Note: Some answer dots will not be used.)

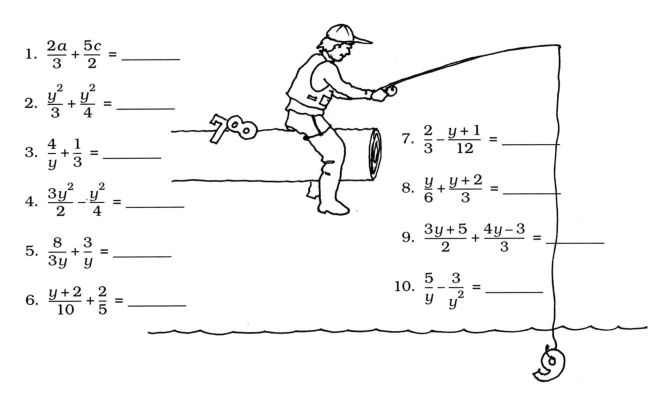

1. $\dfrac{2a}{3} + \dfrac{5c}{2} =$ _____

2. $\dfrac{y^2}{3} + \dfrac{y^2}{4} =$ _____

3. $\dfrac{4}{y} + \dfrac{1}{3} =$ _____

4. $\dfrac{3y^2}{2} - \dfrac{y^2}{4} =$ _____

5. $\dfrac{8}{3y} + \dfrac{3}{y} =$ _____

6. $\dfrac{y+2}{10} + \dfrac{2}{5} =$ _____

7. $\dfrac{2}{3} - \dfrac{y+1}{12} =$ _____

8. $\dfrac{y}{6} + \dfrac{y+2}{3} =$ _____

9. $\dfrac{3y+5}{2} + \dfrac{4y-3}{3} =$ _____

10. $\dfrac{5}{y} - \dfrac{3}{y^2} =$ _____

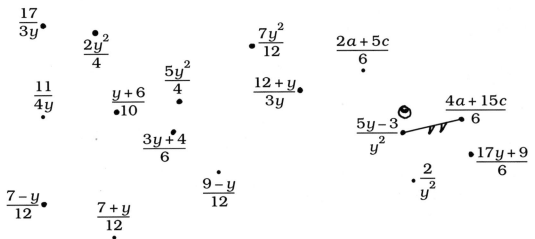

$\dfrac{17}{3y}$

$\dfrac{2y^2}{4}$

$\dfrac{7y^2}{12}$

$\dfrac{2a+5c}{6}$

$\dfrac{11}{4y}$

$\dfrac{y+6}{10}$

$\dfrac{5y^2}{4}$

$\dfrac{12+y}{3y}$

$\dfrac{5y-3}{y^2}$

$\dfrac{4a+15c}{6}$

$\dfrac{3y+4}{6}$

$\dfrac{9-y}{12}$

$\dfrac{2}{y^2}$

$\dfrac{17y+9}{6}$

$\dfrac{7-y}{12}$

$\dfrac{7+y}{12}$

(continued)

© 1983, 2001 J. Weston Walch, Publisher   129   *80 Activities to Make Basic Algebra Easier*

CHAPTER **8**

Graphing and Systems of Linear Equations

This chapter will show students how to graph linear equations with two variables, and some of the ways in which you can solve systems of linear equations.

Teaching Tips

- What is graphing good for? Before you start graphing, put the equation $x + y = 5$ on the board, and ask students for the answer. Most students will answer $2 + 3$, $1 + 4$, etc. Finally, one student will say the magic word—infinity. That is your cue to ask, "How can we show an infinite number of answers?" This question provides a natural lead into graphing linear equations.

- Students seem to appreciate graphing more if you can show them some practical applications. One easy example is to graph the price of a share of stock over a given period of time and then ask students to predict what will happen next to the price of the stock.

Overview of Activities in this Chapter

67. **Ordered Pairs**

   Overview: Introduces using ordered pairs to locate points, with a dot-to-dot practice activity.

68. **Graphing Equations**

   Overview: Introduces creating a table of values for the variables in an equation, then graphing the equation.

Chapter 8: Graphing and Systems of Linear Equations

69. **Solving Systems of Equations Graphically**

    Overview: Introduces finding an ordered pair that makes two equations true at the same time, with a dot-to-dot practice activity.

70. **Solving Systems of Equations (Addition and Multiplication Method)**

    Overview: Introduces the addition and multiplication method of solving systems of equations with two variables.

71. **Solving Systems of Equations (Substitution Method)**

    Overview: Introduces the substitution method of solving systems of equations with two variables, with a cross-number practice activity.

72. **Finding the Slope of a Line**

    Overview: Introduces the concept of slope in algebra.

Name _____

Date _____

Ordered Pairs

Before you can graph equations you must be able to locate points, represented by *ordered pairs*, on a plane. Graph paper is usually used to represent the plane. Start by using a ruler to draw two number lines on graph paper that intersect at zero and form right angles.

The horizontal number line is called the **x** axis, and the vertical number line is called the **y** axis.

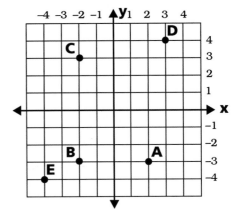

The following examples show how to locate points represented by *ordered pairs* on the plane. Ordered pairs have two numerals inside parentheses. The first numeral represents *x*, the distance to the left or right of the *y* axis. The second value represents *y*, the distance above or below the *x* axis.

Example 1: Locate the point represented by (2, –3).
Start at (0, 0), called the *origin*. Go to the *right* two units and *down* three. (See point *A* above.)

Example 2: Locate the point represented by (–2, –3).
Start at the origin and move to the *left* two and *down* three. (See point *B* above.)

Directions   Use the drawing above to do the following:

Name the ordered pairs that represent:

1. Point C _____

2. Point D _____

3. Point E _____

Remember, in an ordered pair, *x* is the first value and *y* is the second.

Draw in the points represented by these ordered pairs:

4. (–3, 2)

5. (3, –4)

Check your work. Then go on to the puzzle on the next page.

(continued)

Ordered Pairs *(continued)*

It looks as if our friend is in trouble. Plot the points represented by the ordered pairs shown below. If you connect the points in order, you'll see what the problem is.

1. (2, 9)
2. (–2, 7)
3. (–3, 6)
4. (–5, 5)
5. (–6, 4)
6. $(-5\frac{1}{2}, 2\frac{1}{2})$
7. (–5, 4)
8. (–3, 5)
9. (–5, 3)
10. (–2, 4)
11. (0, 3)
12. (–1, 1)
13. (0, 0)
14. (–3, –5)

15. (–1, –3)
16. (–2, –8)
17. (0, –6)
18. (1, –10)
19. (3, –8)
20. (4, –8)
21. (4, –7)
22. (6, –8)
23. (3, –2)
24. (4, –2)
25. (1, 1)
26. (2, 4)
27. (1, 7)
28. (2, 9)

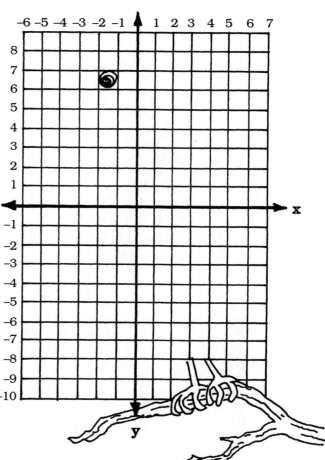

Name _____

Date _____

Graphing Equations

What values for x and y would make this equation true: $x + y = 3$?

If $x = 1$ and $y = 2$,
$1 + 2 = 3$

If $x = 2$ and $y = 1$,
$2 + 1 = 3$

If $x = \frac{1}{2}$ and $y = 2\frac{1}{2}$,
$\frac{1}{2} + 2\frac{1}{2} = 3$

If $x = -2$ and $y = 5$,
$-2 + 5 = 3$

There's an infinite number of values that will work. How can you solve it?

All you can do is make a drawing that represents all the possible answers.

Example 1: Graph $x + y = 3$.

First make up a table of values (ordered pairs) for x and y that will work in the equation.

x	y	
0	3	(If $x = 0$, y must equal 3.)
3	0	(If $y = 0$, x must equal 3.)
1	2	(If $x = 1$, y must equal 2.)
2	1	(If $x = 2$, y must equal 1.)
-1	4	(If $x = -1$, y must equal 4.)

Try to avoid fractions, since they are hard to graph.

Now, graph the points that represent the answers from the table: (0, 3), (3, 0), (1, 2), (2, 1), and (-1, 4). Use a ruler to connect the points.

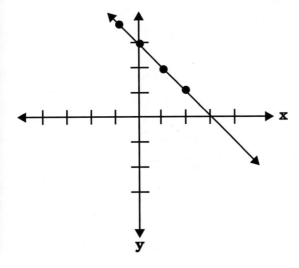

The "line" represents all the answers to the equation.

(continued)

       80 Activities to Make Basic Algebra Easier

Name _____

Date _____

Graphing Equations *(continued)*

Example 2: Graph $y = 2x$.

Start by making up a table of values:

x	y	
0	0	(If $x = 0$, $y = 2 \cdot 0 = 0$.)
1	2	(If $x = 1$, $y = 2 \cdot 1 = 2$.)
2	4	(If $x = 2$, $y = 2 \cdot 2 = 4$.)
–1	–2	(If $x = –1$, $y = 2 \cdot –1 = –2$.)

Now graph the equation.

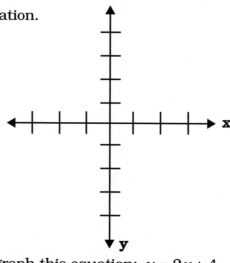

Directions

1. See if you can graph this equation: $y = 2x + 4$

   Fill in this table of values first.

x	y	
0	___	$2 \cdot 0 + 4 =$
___	0	$0 = 2x + 4$ (Solve the equation.)
1	___	$(2 \cdot 1) + 4 =$
2	___	$(2 \cdot 2) + 4 =$
–1	___	$(2 \cdot –1) + 4 =$

Now, graph the equation here.

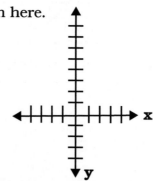

Check your work. Then try the puzzle on the next page.

(continued)

<u>Graphing Equations</u> *(continued)*

Directions   2. Graph each of the following equations. The graph of each equation should pass through a letter. Write down the letter that corresponds to each equation in the spaces shown below. If your work is correct, a four-word message will appear.

1. $x + y = 6$

2. $y = 2x$

3. $y = x + 2$

4. $y = 2x + 4$

5. $-2x = y$

6. $x + y = -6$

7. $y = -2x + 4$

8. $y = 3x + 6$

9. $y = \frac{1}{3}x - 2$

10. $4y + 2x = -8$

11. $2y + x = -10$

12. $x = -1$

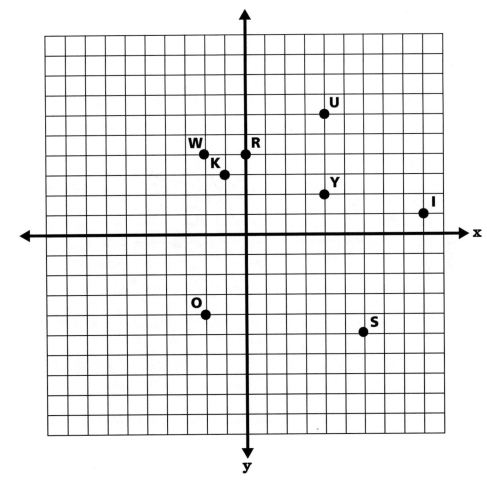

List the letters below:

___ ___ ___ ___   ___ ___ ___ ___   ___ ___ ___ ___
(1) (2) (3) (4)   (5) (6) (7) (8)   (9) (10) (11)(12)

             80 Activities to Make Basic Algebra Easier

ACTIVITY 69

Solving Systems of Equations Graphically

How can you find values for *x* and *y* that make both equations true at the same time? If you are given two equations in *x* and *y*, they can be solved (an ordered pair can be found that works for both equations) by graphing them both on the same *x* and *y* axis.

Example: Solve $x + y = 3$
$$y = 2x$$

Make up a table for each equation.

$x + y = 3$ $\qquad\qquad$ $y = 2x$

x	y
0	3
3	0
1	2
2	1

x	y
0	0
1	2
2	4

Graph the two equations and look for a point of intersection.

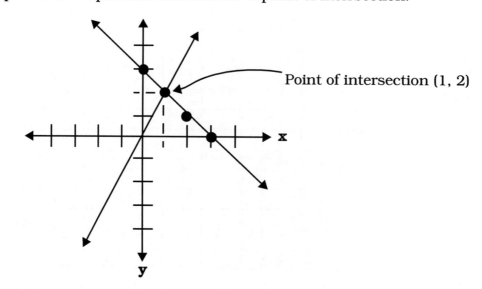

Point of intersection (1, 2)

Check your answer by substituting *x* = 1 and *y* = 2 in the original equations.

$x + y = 3$ $\qquad\qquad$ $y = 2x$
$1 + 2 = 3$ $\qquad\qquad$ $2 = 2 \cdot 1$

Directions Now, try the puzzle on page 139.

Now, try the puzzle on page 139.

Name _____

Date _____

Solving Systems of Equations Graphically (continued)

Directions Solve the following equations by graphing on a separate sheet of graph paper. Show your work. Then, on the graph at the bottom of this page, graph the ordered pairs that represent the answers. Connect the dots in the same order as the problems are numbered and a cartoon face will appear.

1. $x + y = 8$
   $y - x = 2$

2. $x + y = 5$
   $x - y = 1$

3. $x + y = 3$
   $x = -4y$

4. $y = -3x$
   $x + y = -2$

5. $2x + y = 2$
   $x + y = -1$

6. $3x + 2y = -17$
   $y = x - 1$

7. $y = 2x + 10$
   $x + y = 1$

8. $3x - y = 4$
   $x - y = -2$

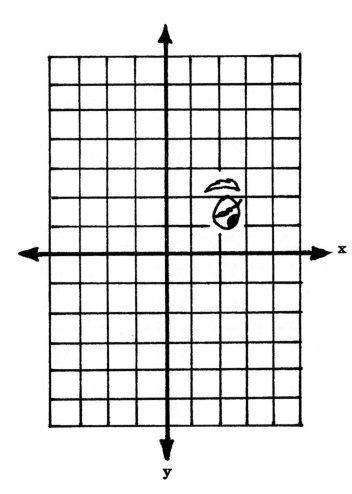

Name _____

Date _____

Solving Systems of Equations
(Addition and Multiplication Method)

> Look what happens when we add both sides of these arithmetic problems together.

$$3 + 2 = 5$$
$$\underline{3 - 2 = 1}$$
$$6 \quad\ = 6$$

The same logic leads to a method that can be used to solve *some* equations with two variables.

Example 1: Solve for x and y.

$$x + y = 5$$
$$\underline{x - y = 1}$$
$$2x \quad\ = 6$$
$$x = 3$$

> Add the two equations together. Note that one of the variables, in this case, y, drops out.

Go back and solve for y by substituting 3 in place of x in either equation.

$$3 + y = 5$$
$$y = 2$$

> Put 3 in place of x in the first equation.

$$3 - 2 = 1$$

> Check that *both* values work in the second equation.

Example 2:

> Add the equations together.

$$2x + \ y = 7$$
$$\underline{x + \ y = 4}$$
$$3x + 2y = 11$$

> What happened? Neither variable dropped out.

Since multiplying *both* sides of an equation by the same number results in an equivalent equation, try multiplying both sides of one of the equations by negative one.

$$2x + \quad\ y = \quad\ 7$$
$$\underline{(-1)x + (-1)\,y = (-1)4}$$
$$x = 3$$

> (Multiply *each* term by –1. Then, add the equations.)

(continued)

Name _____

Date _____

Solving Systems of Equations
(Addition and Multiplication Method) *(continued)*

Directions   Solve the following equations for *x* and *y*. Place the *x* answer *horizontally* in the squares to the right of the problems. If your work is correct, the corresponding *y* answer will appear in the vertical columns. (The first problem has been done for you.)

1. $x + y = 498$

   $x - y = 14$

2. $2x - y = 298$

   $x + y = 911$

3. $2y - x = 984$

   $y - x = 350$

4. $x + y = 279$

   $3x + y = 551$

5. $2x + 3y = 1,830$

   $2x - y = 518$

6. $x + 2y = 1,640$

   $2x + 2y = 2,020$

7. $2x + y = 436$

   $3x - 2y = 192$

8. $x + y = 850$

   $2x + 3y = 2,226$

9. $2x + 3y = 1,270$

   $3x - 2y = 306$

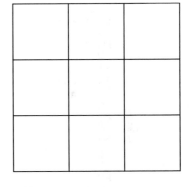

1. $y =$	2. $y =$	3. $y =$
2	**5**	**6**

1. $x =$

2. $x =$

3. $x =$

4. $y =$	5. $y =$	6. $y =$

4. $x =$

5. $x =$

6. $x =$

7. $y =$	8. $y =$	9. $y =$

7. $x =$

8. $x =$

9. $x =$

         80 Activities to Make Basic Algebra Easier

Name _____

Date _____

Solving Systems of Equations (Substitution Method)

Example 1:

$$y = 2x$$
$$x + y = 12$$
$$x + 2x = 12$$

> In this problem $2x$ is another name for y. So we can *substitute* it for y in the *second* equation.

Combine like terms:
$$3x = 12$$

Divide by 3:
$$x = 4$$

Solve for y:
$$y = 2 \cdot 4$$ (Use the first equation.)
$$y = 8$$

Check:
$$4 + 8 = 12$$ (Use the second equation.)

Example 2:

$$2x + y = 13$$
$$x = y + 2$$

> The second equation tells us that $y + 2$ is another name for x.

Substitute $y + 2$ for x in the first equation:
$$2(y + 2) + y = 13$$

Multiply $2(y + 2)$:
$$2y + 4 + y = 13$$

Combine like terms:
$$3y + 4 = 13$$

Subtract 4:
$$3y = 9$$

Divide by 3:
$$y = 3$$

Solve for x:
$$x = y + 2 \quad \text{(second equation)}$$
$$x = 3 + 2$$
$$x = 5$$

Check:
$$2x + y = 13 \quad \text{(first equation)}$$
$$(2 \cdot 5) + 3 = 13$$

Sometimes you have to rewrite an equation in order to use this method.

Example 3:

$$x + y = 9$$
$$x - y = 3$$
$$x - y + y = y + 3$$
$$x = y + 3$$

> Add y to *both* sides of the second equation.

Put $y + 3$ in place of x in the *first* equation.

$$(y + 3) + y = 9$$
$$2y + 3 = 9$$
$$2y = 6$$
$$y = 3$$

(continued)

       80 Activities to Make Basic Algebra Easier

Name _____

Date _____

Solving Systems of Equations
(Substitution Method) *(continued)*

Directions Solve the equations below for x and y. Then place the y values in the squares of the cross-number puzzle at the bottom of the page.

Across

1. $x + y = 25$ $x =$ _____

    $y = 4x$ $y =$ _____

2. $x = y + 8$ $x =$ _____

    $2x + y = 52$ $y =$ _____

3. $3x = y$ $x =$ _____

    $2x = y - 10$ $y =$ _____

5. $3x + y = 33$ $x =$ _____

    $x = y - 13$ $y =$ _____

7. $x - y = -10$ $x =$ _____

    $3x = y + 18$ $y =$ _____

Down

1. $x + y = 34$ $x =$ _____

    $y = x + 14$ $y =$ _____

2. $y = 5x$ $x =$ _____

    $6x - y = 2$ $y =$ _____

4. $2x - y = 6$ $x =$ _____

    $x - y = -4$ $y =$ _____

6. $y - 3x = 22$ $x =$ _____

    $x + y = 102$ $y =$ _____

7. $x + y = 30$ $x =$ _____

    $y = x + 16$ $y =$ _____

Finding the Slope of a Line

Name _____

Date _____

The **grade** or **steepness** of a road is usually expressed as a percent. A 30% grade means that for every 100 feet of horizontal distance, the road **rises** 30 feet.

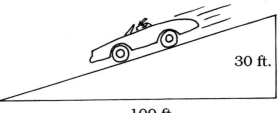

30 ft.

100 ft.

Danger: 30% grade

$$\text{Grade} = \frac{\text{Rise}}{\text{Run}} = \frac{30}{100} = 30\%$$

In algebra, the steepness of a line in a graph is called slope.

$$\text{Slope} = \frac{\text{Rise}}{\text{Run}} \quad \text{or} \quad \frac{\text{Vertical change}}{\text{Horizontal change}}$$

Example: Find the slope of a line containing the points (1, 4) and (3, 1).

Step 1: First graph the line.
There are several ways to find the slope by counting.

Step 2: Starting at (1, 4), count *down* to get a rise of *negative* 3. Then count to the *right* until you get to the point (3, 1). This gives you a run of +2.

$$\text{Slope} = \frac{\text{Rise}}{\text{Run}} = \frac{-3}{+2} = -\frac{3}{2}$$

Step 3: Starting at (3, 1), count *up* to get a rise of +3. Then count to the *left* to get a run of –2.

$$\text{Slope} = \frac{\text{Rise}}{\text{Run}} = \frac{+3}{-2} = -\frac{3}{2}$$

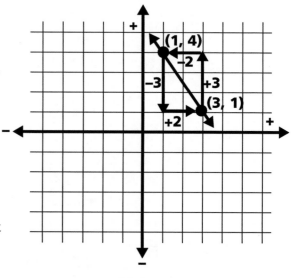

(continued)

           80 Activities to Make Basic Algebra Easier

ACTIVITY 72

Finding the Slope of a Line *(continued)*

Directions   Graph the following ordered pairs and find the slope.

1. (1, –2), (–3, 4)

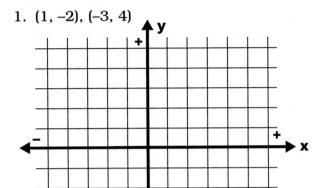

Slope = _____

2. (3, 2), (4, –2)

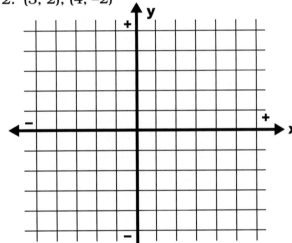

Slope = _____

3. (1,1), (–3, –2)

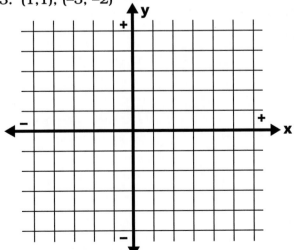

Slope = _____

(continued)

ACTIVITY 72

Finding the Slope of a Line *(continued)*

Here's a formula you can use to find the slope of a line without graphing. Look at the ordered pairs in the previous example: (1, 4) and (3, 1). Think of the first ordered pair as (x_1, y_1) and the second ordered pair as (x_2, y_2). Substitute the x and y values in the formula shown below.

$$\text{Slope} = \frac{y_2 - y_1}{x_2 - x_1} = \frac{1 - 4}{3 - 1} = \frac{1 + -4}{2} = \frac{-3}{+2} = -\frac{3}{2}$$

Directions   Now try problems 4–6 using the formula.

4. (1, –2), (–3, 4)   5. (3, 2), (4, –2)   6. (1, 1), (–3, –2)

$$\text{Slope} = \frac{4 - (-2)}{-3 - 1}$$

Slope = _____   Slope = _____   Slope = _____

7. Find the slope for (3, 2) and (5, 2). Slope = _____

8. Describe the slope for (2, 3) and (2, 5). _____

_____

_____

CHAPTER 9

Rational and Irrational Numbers

This chapter will show students how to solve problems that involve irrational numbers.

Teaching Tips

- Do not be too quick to let your students use a calculator for square roots. Have them try some problems first using estimates or a divide-and-average approach. Use a geometric approach to have students construct square roots (see Activity 76).

- Start the Pythagorean Theorem with a manipulative approach (see Activity 76), not formal proof.

- When you feel students are ready for a proof of the Pythagorean Theorem, start with one that is mainly manipulative (see Activity 80), and then move to a more formal approach.

- Discuss practical applications of 3-4-5 triangles and building construction. The 3-4-5 triangle was used by ancient craftsmen to ensure that the corners of a building would be square (90°). Here's how it works.

  1. Lay out the corner of a building using string and stakes in the ground (see diagram that follows);

  2. Mark off 3 feet on one side of the corner and 4 feet on the other side;

  3. Stretch the tape measure between the 3 foot and 4 foot marks forming a triangle; and

  4. If the distance between the 3 foot and 4 foot marks is exactly 5 feet, you have a right angle. The reason this works is that a 3-4-5 *right* triangle has been formed ($3^2 + 4^2 = 5^2$). If the hypotenuse of the triangle is not 5 feet, adjust the stakes until it is. This is a great outdoor group activity.

Chapter 9: Rational and Irrational Numbers

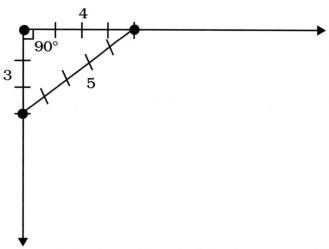

Note: Even though this idea is over 2,000 years old, it is still used by craftsmen today. See Activity 75 for a related activity.

Overview of Activities in this Chapter

73. **Square Root (Perfect Squares)**

    Overview: Introduces the guess-and-average approach to finding square roots, with a cross-number practice activity.

74. **Square Root (Irrational Numbers)**

    Overview: Introduces square roots of irrational numbers, with a dot-to-dot practice activity.

75. **An Introduction to the Pythagorean Theorem**

    Materials: Graph paper, scissors

    Overview: Introduces the Pythagorean Theorem through a physical model.

76. **Square Root Approximation (Pythagorean Theorem)**

    Materials: Metric ruler, square root table

    Overview: Shows students how to use measurements and the Pythagorean Theorem to approximate square roots.

77. **A Pattern for Completing the Square**

    Overview: Introduces some more approaches to solving quadratic equations.

Teacher Guide Page

Chapter 9: Rational and Irrational Numbers

78. **The Quadratic Formula**

    Overview: Introduces the quadratic formula, with a dot-to-dot practice activity.

79. **A Pythagorean Experiment**

    Materials: Index cards, scissors, paper

    Overview: Students create a physical model of the Pythagorean Theorem, then use the model to explain how the models prove that, in a right triangle, $a^2 + b^2 = c^2$.

80. **Finding the Distance Between Two Points on a Graph**

    Overview: Students learn how to find the distance between two points by graphing, then use a number pattern to develop a shortcut to finding the distance.

Name _____

Date _____

Square Root (Perfect Squares)

The opposite (inverse) of the square of a number is called the *square root*. ($\sqrt{\ }$ is the square root symbol.)

> **Examples:**
> $3^2 = 3 \times 3 = 9$, $\sqrt{9} = 3$ (Read as "The square root of 9 is 3.")
> $5^2 = 25$, $\sqrt{25} = 5$ (Read as "The square root of 25 is 5.")

Directions   Write the square roots of the following:

1. $\sqrt{36}$ = _____   4. $\sqrt{81}$ = _____

2. $\sqrt{49}$ = _____   5. $\sqrt{144}$ = _____

3. $\sqrt{100}$ = _____   6, $\sqrt{1}$ = _____

Sometimes it's difficult to guess the square root of a number.

Example: Find $\sqrt{2304}$

Start by making a guess as to what the answer could be.

I don't know. I'll guess 35.

OK, now divide 35 into 2304. If you're correct, your quotient should be 35.

I told you I couldn't do it! 65 isn't close to 35.

$$\begin{array}{r} 65 \\ 35{\overline{)2304}} \\ 210 \\ \hline 204 \\ 175 \\ \hline 29 \end{array}$$

Don't worry about it. You now know that your answer is *between* 35 and 65.

To find a number between 35 and 65, "average" the two numbers (add them together and divide by 2).

$$\begin{array}{r} 65 \\ + 35 \\ \hline 100 \end{array} \qquad \begin{array}{r} 50 \\ 2{\overline{)100}} \end{array} \qquad \text{Try 50 as an answer.}$$

(continued)

80 Activities to Make Basic Algebra Easier

Name _____

Date _____

Square Root (Perfect Squares) *(continued)*

Divide 50 into 2304.

It still didn't work. I'll never get it.

$$\begin{array}{r} 46 \\ 50)\overline{2304} \\ 200 \\ \hline 304 \\ 300 \\ \hline 4 \end{array}$$

Don't give up! You now know that your answer is between 46 and 50.

"Average" 46 and 50.

$$\begin{array}{r} 46 \\ +\ 50 \\ \hline 96 \end{array}$$

$$\begin{array}{r} 48 \\ 2)\overline{96} \\ 8 \\ \hline 16 \\ 16 \end{array}$$

Divide 48 into 2304.

It worked! I always knew I could do it.

$$\begin{array}{r} 48 \\ 48)\overline{2304} \\ 192 \\ \hline 384 \\ 384 \end{array}$$

The closer your first guess, the easier your work will be.

Use what you have learned to solve the following problems. Place your answers in the cross-number puzzle on the next page.

Across

1. $\sqrt{169}$ = _____

2. $\sqrt{676}$ = _____

6. $\sqrt{1849}$ = _____

8. $\sqrt{5776}$ = _____

10. $\sqrt{13,924}$ = _____

11. $\sqrt{3364}$ = _____

12. $\sqrt{3136}$ = _____

Down

1. $\sqrt{361}$ = _____

3. $\sqrt{7396}$ = _____

4, $\sqrt{4096}$ = _____

5. $\sqrt{1156}$ = _____

7. $\sqrt{1225}$ = _____

9. $\sqrt{9604}$ = _____

12. $\sqrt{3481}$ = _____

(continued)

Name _____

Date _____

Square Root (Perfect Squares) *(continued)*

Square Root (Irrational Numbers)

Example: Find $\sqrt{34}$

$5^2 = 25$
$6^2 = 36$
It doesn't work!

Numbers like $\sqrt{34}$ are called *irrational* numbers. We can approximate them using decimals.

Irrational numbers are named by decimals that do not terminate and do not repeat. (**Example:** .1313313331 . . .)

Your first guess is between 5 and 6, which is 5.5.

Divide 34 by 5.5

```
          6.18
   5.5 ) 34.0.00
          33 0
           1 00
             55
            450
            440
             10
```

(This time, carry your answer out an extra decimal place to the hundredths.)

Your second guess is between 5.5 and 6.18. Average 5.5 and 6.18.

```
   5.50
 + 6.18
  11.68
```

```
       5.84
  2 ) 11.68
       10
        16
        16
         8
         8
```

Now divide 34 by 5.84.

```
                5.821
  5.84 ) 34.00.000
          29 20
           4 800
           4 672
            1280
            1168
            1120
             584
             536
```

You can see that this process can be continued forever. At this point we can assume that the answer is approximately 5.83 to the nearest hundredth.

Now see if you can work the puzzle on page 154.

(continued)

         80 Activities to Make Basic Algebra Easier

Name _____

Date _____

Square Root (Irrational Numbers) *(continued)*

Directions   Use what you have learned about working with square roots to solve the following problems. (Round each answer to the nearest *tenth*.) Locate the answers below. If you connect the dots below each correct answer in the same order in which the problems are numbered, a drawing will appear that represents one of the first irrational numbers that you studied. (Note: Some answer dots will not be used.)

1. $\sqrt{13}$ = _____

2. $\sqrt{21}$ = _____

3. $\sqrt{87}$ = _____

4. $\sqrt{130}$ = _____

5. $\sqrt{99}$ = _____

6. $\sqrt{135}$ = _____

7. $\sqrt{150}$ = _____

8. $\sqrt{236}$ = _____

9, $\sqrt{328}$ = _____

10. $\sqrt{19}$ = _____

11. $\sqrt{32}$ = _____

12. $\sqrt{3.7}$ = _____

13. $\sqrt{12.8}$ = _____

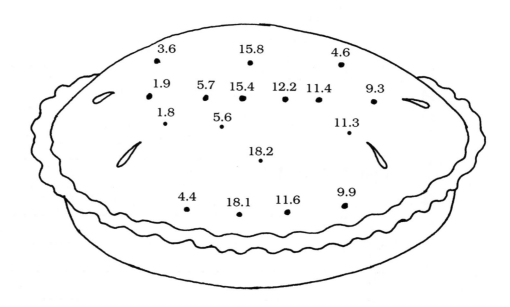

Give a decimal approximation for the number represented by the drawing (nearest hundredth). _____

An Introduction to the Pythagorean Theorem

Name _____

Date _____

Directions For this activity you will need a sheet of graph paper and scissors. Use the scissors to cut out graph paper strips with lengths of 3, 4, and 5 units (see below).

3 ☐☐☐

4 ☐☐☐☐

5 ☐☐☐☐☐

Directions Use the strips to form a *right* triangle.

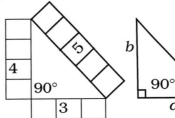

> In any *right* triangle, the sides that form the right angle (*a* and *b*) are called the *legs*. . .

> . . . and the third side, *c*, is called the *hypotenuse.*

Directions Cut out some more graph paper strips with the following lengths: 6, 7, 8, 10, 12, 13, 16, and 20.

Directions Try to form a right triangle using 5, 12, and 13. Record your results in the table below. Use the same procedure for the rest of the table. The first triangle has been done for you.

> Not all of the triangles will be right triangles. Check with a protractor if you're not sure.

Length of sides							Is a right triangle formed?
a	b	c	a^2	b^2	$a^2 + b^2$	c^2	
3	4	5	9	16	9 + 16 = 25	25	yes
5	12	13					
4	5	7					
6	8	10					
12	16	20					
6	10	13					

Study the completed table. It looks as if in a *right* triangle $a^2 + b^2 =$ _____.

(continued)

ACTIVITY 75

An Introduction to the Pythagorean Theorem *(continued)*

$a^2 + b^2 = c^2$ is a way of symbolizing the Pythagorean theorem. This is one of the most important concepts in mathematics. It can be used in a number of different ways. Remember, it only applies to *right* triangles.

Example 1: You are told that the legs (a and b) of a right triangle are 6 and 8 centimeters long. Use the Pythagorean theorem to find the hypotenuse (c).

First, make a drawing to help your thinking.

$$\text{If } a^2 + b^2 = c^2,$$
$$\text{then } 6^2 + 8^2 = c^2.$$
$$36 + 64 = c^2$$
$$c^2 = 100$$
$$c = \sqrt{100} = 10$$

Answer: 10 centimeters

Example 2: Sometimes one of the legs is missing. Find b, if $a = 3$ cm and $c = 5$ cm.

$$a^2 + b^2 = c^2$$
$$3^2 + b^2 = 5^2$$
$$9 + b^2 = 25$$
$$b^2 = 25 - 9 = 16$$
$$b = \sqrt{16} = 4 \text{ cm}$$

(continued)

ACTIVITY 75

An Introduction to the
Pythagorean Theorem *(continued)*

1. The lengths of two of the sides of two right triangles are shown below. Use the Pythagorean theorem to find the missing sides. Show your work.

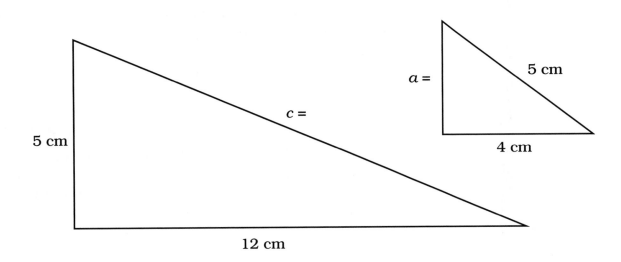

Check your answer by measuring the missing side in centimeters.

         80 Activities to Make Basic Algebra Easier

Name _____

Date _____

Square Root Approximation (Pythagorean Theorem)

For this activity you will need a metric ruler and a square root table.

Example: Make a drawing that represents $\sqrt{5}$.

(a) Use the metric ruler to draw a *right* triangle with legs 1 and 2 centimeters in length.

2 | | c

1

(b) The Pythagorean theorem tells us that $1^2 + 2^2 = c^2$ and that $c = \sqrt{5}$. Use the metric ruler to measure the length of c to the nearest tenth of a centimeter. Find an approximation for $\sqrt{5}$ in your table and round it to the nearest tenth. Compare your measurement to the approximation from the table. Did you get 2.2?

Directions Use the metric ruler to measure c in each of the triangles shown below (nearest tenth). Check each answer using the Pythagorean theorem and a square root table. Show your work. The first one is done for you.

1.

1 cm | | $c \approx 1.4$ cm

1 cm

$1^2 + 1^2 = c^2$
$c = \sqrt{2}, \ \sqrt{2} \approx 1.4$

2.

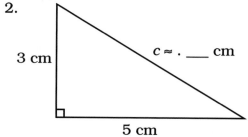

3 cm

$c \approx .$ ___ cm

5 cm

3.

1 cm | | $c \approx$ ___ cm

3 cm

4.

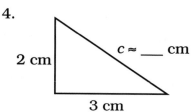

2 cm

$c \approx$ ___ cm

3 cm

Directions Use the Pythagorean theorem to help you make drawings of right triangles where the hypotenuse represents the following square roots:

5. $\sqrt{20}$ 6. $\sqrt{18}$ 7. $\sqrt{41}$

Directions Check your work using a metric ruler and square root table as you did in problems 1–4.

Name _____

Date _____

A Pattern for Completing the Square

You have already learned how to solve quadratic equations by factoring. There are, however, other ways. Some of these methods will be developed in this activity.

Example 1: Solve: $y^2 = 25$

Take the square root of *both* sides.

$y = \sqrt{25}$

$y = +5 \text{ or } -5$

> Note that there are two possible answers: $(-5)^2$ also equals 25.

Example 2: Solve: $x^2 - 36 = 0$

This equation could be solved by factoring or by taking the square root of both sides.

> First, add 36 to both sides.

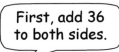

$x^2 - 36 + 36 = 0 + 36$

$x^2 = 36$

$x = \sqrt{36}$

> Now take the square root of both sides.

$x = \pm 6$

> Read as "x equals positive or negative 6."

Example 3: Solve: $y^2 - 5 = 0$

Add 5 to each side: $y^2 = 5$

Take the square root of both sides: $y = \pm \sqrt{5}$

Example 4: Solve: $(y - 3)^2 = 25$

Take the square root of both sides: $y - 3 = \pm 5$

$y - 3 = +5 \quad \text{and} \quad y - 3 = -5$

Add 3 to each side: $y - 3 + 3 = 5 + 3 \qquad y - 3 + 3 = -5 + 3$

$y = 8 \qquad\qquad y = -2$

Check: $(8 - 3)^2 = 25 \qquad (-2 - 3)^2 = 25$

Directions Use what you have learned to solve the following. Show your work.

1. $y^2 - 49 = 0$

2. $(n - 4)^2 = 36$

3. $(x + 3)^2 = 16$

Check your work before you go on to page 160.

(continued)

A Pattern for Completing the Square *(continued)*

Name _____

Date _____

Directions A typical quadratic equation can be symbolized by $ax^2 + bx + c = 0$. Fill in the missing values in the table below. (The first two have been done for you.)

Equation	a	b	c	$\frac{1}{2}b$	$(\frac{1}{2}b)^2$	Factored Form of Equation
$x^2 + 6x + 9 = 0$	1	6	9	3	$3^2 = 9$	$(x + 3)(x + 3) = 0$ *or* $(x + 3)^2 = 0$
$x^2 - 8y + 16 = 0$	1	–8	16	–4	$(-4)^2 = 16$	$(a - 4)(a - 4) = 0$ *or* $(a - 4)^2 = 0$
$x^2 + 8x + 16 = 0$						
$n^2 - 6n + 9 = 0$						
$y^2 + 2y + 1 = 0$						

Directions Use what you have written in the table to answer these questions.

4. In each of the equations $a =$ _____.

5. In each of the equations $(\frac{1}{2}b)^2$ equals what other value in the table? _____

6. What do you notice about the factored form of each equation? _____ _____

Now let's put together what was in the first part of the activity with the patterns in the table. This leads to a new method for solving quadratic equations called "completing the square."

Example: Solve: $y^2 + 8y + 15 = 0$ by completing the square.
Add –15 to each side. $\qquad\qquad y^2 + 8y = -15$.

> If the left side of the equation was a perfect square, we could take the square root of both sides, as we did before.

> The table shows that the constant (c) equals $(\frac{1}{2}b)^2$ in a perfect trinomial square.

Add $(\frac{1}{2} \cdot 8)^2$ to both sides. $\qquad y^2 + 8y + 16 = -15 + 16$
$y^2 + 8y + 16$ is a perfect trinomial square: $\quad (y + 4)^2 = 1$
Take the square root of both sides. $\qquad y + 4 = \pm 1$

$$y + 4 = +1 \qquad and \qquad y + 4 = -1$$
$$y + 4 - 4 = 1 - 4 \qquad\qquad y + 4 - 4 = -1 - 4$$
$$y = -3 \qquad\qquad\qquad y = -5$$

ACTIVITY 78

The Quadratic Formula

$x = \dfrac{-b \pm \sqrt{b^2 - 4ac}}{2a}$ is called the **quadratic formula**. It can be used to solve any quadratic equation of the form $ax^2 + bx + c = 0$.

Example 1: Use the quadratic formula to solve $x^2 + 4x - 12 = 0$.

First, determine what a, b, and c are.

$$ax^2 + bx + c = 0$$
$$x^2 + 4x - 12 = 0$$

You can see that $a = 1$, $b = 4$, and $c = -12$.

Now substitute these values in the formula:

$$\dfrac{-b \pm \sqrt{b^2 - 4ac}}{2a}$$

$$\dfrac{-4 \pm \sqrt{4^2 - 4(1)(-12)}}{2(1)} = \dfrac{-4 \pm \sqrt{16 + 48}}{2}$$

$$= \dfrac{-4 \pm \sqrt{64}}{2} = \dfrac{-4 \pm 8}{2}$$

$$\dfrac{-4 + 8}{2} = \dfrac{4}{2} = 2, \quad \dfrac{-4 - 8}{2} = \dfrac{-12}{2} = -6$$

Example 2: $x^2 + 2x - 10 = 0$

Determine values for a, b, and c ($a = 1$, $b = +2$, and $c = -10$).

Substitute the values into the formula:

$$\dfrac{-(2) \pm \sqrt{(2)^2 - 4(1)(-10)}}{2(1)} = \dfrac{-2 \pm \sqrt{4 + 40}}{2}$$

$$= \dfrac{-2 \pm \sqrt{44}}{2} = \dfrac{-2 \pm 2\sqrt{11}}{2} \quad \boxed{\sqrt{44} = \sqrt{4 \cdot 11} = 2 \cdot \sqrt{11}}$$

$$= \dfrac{2(-1 \pm 1\sqrt{11})}{2} = -1 \pm \sqrt{11}$$

(continued)

     80 Activities to Make Basic Algebra Easier

Name _____

Date _____

The Quadratic Formula *(continued)*

> **Example 3:** $x^2 - 8x = -12$
>
> First, arrange the equation in the form $ax^2 + bx + c = 0$.
> (Add 12 to each side.) $x^2 - 8x + 12 = 0$
>
> $a = 1$, $b = -8$, $c = +12$, $\dfrac{-(-8) \pm \sqrt{(-8)^2 - 4(1)(12)}}{2(1)}$
>
> $= \dfrac{+8 \pm \sqrt{64 - 48}}{2} = \dfrac{8 \pm \sqrt{16}}{2}$
>
> $\dfrac{8 + 4}{2} = 6$, $\dfrac{8 - 4}{2} = 2$

(continued)

Name _____

Date _____

The Quadratic Formula *(continued)*

Directions   Dr. Jekyll made a mistake and drank the wrong formula. To find out what happened, use the *quadratic formula* to solve the following problems. Locate the solutions below. If you connect the dots next to each correct solution in the same order in which the problems are numbered, a cartoon will appear.

1. $x^2 + 5x + 6 = 0$
2. $x^2 + x - 6 = 0$
3. $2x^2 + 7x + 3 = 0$
4. $x^2 + 2x - 15 = 0$
5. $x^2 - 7x + 10 = 0$
6. $3x^2 + x - 2 = 0$
7. $x^2 - 7x + 12 = 0$
8. $2x^2 - 7x + 3 = 0$
9. $2x^2 - x - 6 = 0$
10. $3x^2 - 4x - 4 = 0$
11. $x^2 + 6x + 9 = 0$
12. $x^2 - 3x = 10$

13. $2x^2 - 9x + 4 = 0$
14. $2x^2 + 9x = -4$
15. $x^2 - 6x + 9 = 0$
16. $x^2 - 3x + 1 = 0$
17. $x^2 + 3x = 0$
18. $x^2 - 1 = 0$
19. $x^2 - 3x = 0$
20. $(x + 5)^2 = 0$
21. $x^2 - x - 3 = 0$
22. $x^2 - x - 1 = 0$
23. $x^2 - 16 = 0$
24. $x^2 - 4x + 1 = 0$

$\left(\frac{1}{2}, 3\right)\bullet$

$\left(-\frac{2}{3}, 2\right)$

$(5, -2)$

$\left(\frac{2}{3}, -1\right)$

$\dfrac{-b \pm \sqrt{b^2 - 4ac}}{2a}$

$\left(\frac{1}{2}, 4\right)$ -3 $\left(-\frac{3}{2}, 2\right)$

$\bullet (4, 3)$

$\left(-\frac{1}{2}, -4\right)\bullet$

$(5, 2)\bullet$

3

$\left(-\frac{1}{2}, -3\right)$

$\dfrac{3 \pm \sqrt{5}}{2}\bullet$

± 1 $\bullet (0, 3)$ $\dfrac{1 \pm \sqrt{13}}{2}$

$(3, -5)$

$(0, -3)$

-5

$\dfrac{1 \pm \sqrt{5}}{2}\bullet$ ± 4 $(2, -3)$

$2 \pm \sqrt{3}$ $\bullet (-3, -2)$

ACTIVITY 79

Name _____

Date _____

A Pythagorean Experiment

There are many ways to show that the Pythagorean theorem works for all right triangles. The following experiment illustrates one of them.

Materials Needed	
• four index cards	• scissors
• ruler	• plain paper

1. Start by cutting off part of an index card so that a right triangle is formed. (See below.)

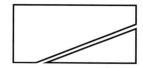

2. Use this triangle as a pattern to make three more cardboard right triangles that are the same size as the first one.

3. Label the legs of each triangle a and b, and the hypotenuse c.

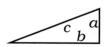

4. Arrange the four triangles as shown below on a separate piece of paper (Figure 1). Use a ruler to extend the sides so that the drawing looks like Figure 2.

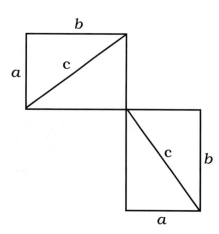

Figure 1

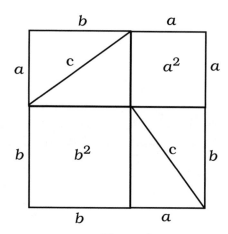

Figure 2

5. Label the areas of the two small squares a^2 and b^2 as shown in Figure 2.

(continued)

         80 Activities to Make Basic Algebra Easier

A Pythagorean Experiment *(continued)*

6. Now use the same four triangles to make a drawing next to the first one that looks like this. (Label the drawing as shown below.)

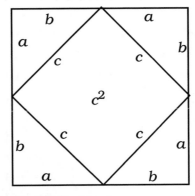

7. Cut out both of your drawings and compare the size of each *large* square. They should be the same (the area of each large square is $(a + b)^2$).

8. Finally, the last step. Carefully tear the four right triangles out of each drawing. Your work should now look like this.

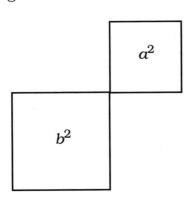 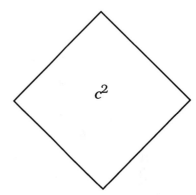

See if you can use the models that you have made to explain to your teacher how this proves that in a right triangle $a^2 + b^2 = c^2$.

I don't understand this!

You can do it. Go back over each step with the models and think about it.

             80 Activities to Make Basic Algebra Easier

ACTIVITY 80

Finding the Distance Between
Two Points on a Graph

(x_1, y_1) (x_2, y_2)

Example: Find the distance between (–2, 1) and (2, 4).

   Step 1: Graph the ordered pair and connect the two points.

   Step 2: Draw segments from the points to form a *right* triangle.

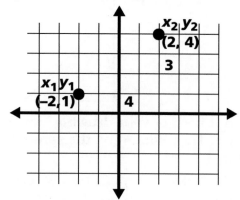

   Step 3: Use the Pythagorean Theorem to find the distance between
   the two points:

$$c^2 = a^2 + b^2$$
$$c^2 = 3^2 + 4^2$$
$$c^2 = 9 \quad + 16$$
$$c = \sqrt{25}$$
$$c = 5$$

(continued)

166   *80 Activities to Make Basic Algebra Easier*

Name _____

Date _____

ACTIVITY **80**

Finding the Distance Between
Two Points on a Graph *(continued)*

Directions   Solve the following problems. Graph the points. Then use the Pythagorean theorem to find the distance between them. Round your answers to the nearest hundredth as needed.

1. (5, 3), (–3, –3)

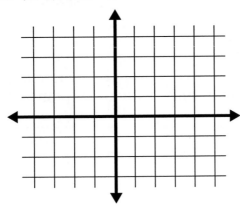

2. (–2, 2), (2, –2)

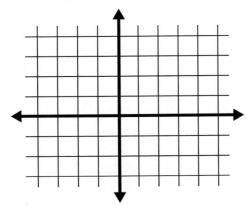

3. (1, 4), (3, 1)

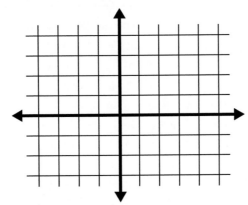

Check your answers. Then go on to page 168.

(continued)

Name _____

Date _____

Finding the Distance Between Two Points on a Graph (continued)

Graphing to find the distance between two points takes time. Let's see if we can use a number pattern to find a shortcut. Start with the ordered pairs from the example on the previous page: (–2, 1) and (2, 4). Think of the first ordered pair as (x_1, y_1) and the second as (x_2, y_2). Finish the problem shown below. (Follow the rules for subtraction of integers.)

$x_2 - x_1 = +2 - -2 = +2 + +2 =$ _____

$y_2 - y_1 = +4 - +1 = +4 + -1 =$ _____

Now look at the legs of the right triangle. Describe in words how you can find the legs of the right triangle without graphing.

_____

_____

Directions   Now try using this approach for the three problems you worked on the previous page.

4. (5, 3), (–3, –3)   5. (—2, 2), (2, –2)   6. (1, 4), (3, 1)

   $x_2 - x_1 =$ _____ $x_2 - x_1 =$ _____ $x_2 - x_1 =$ _____

   $y_2 - y_1 =$ _____ $y_2 - y_1 =$ _____ $y_2 - y_1 =$ _____

$d = \sqrt{(x_2 - x_1)^2 + (y_2 - y_1)^2}$ is called the **distance formula**.

Directions   See if you can explain how the formula is derived.

7. $x_2 - x_1$ gives the _____

8. $y_2 - y_1$ gives the _____

9. How is the Pythagorean theorem used in the formula?

_____

_____

10. See if you can solve problems 1–3 using the formula.

Answer Key

Activity 1: Order of Operations

1. 19
2. 35
3. 15
4. 30
5. 22
6. 0
7. 9
8. 10

Activity 2: Basic Exponents I

2. $6^2 = 36$
3. $5^2 = 25$
4. $10^2 = 100$
5. $8^2 = 64$
6. 81
7. 144
8. 49
9. 1
10.

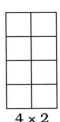 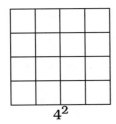

4 × 2 4^2

Activity 3: Basic Exponents II

1.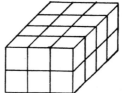

2.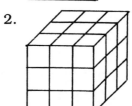

3.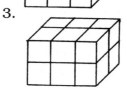

$3 \times 3 \times 3 \times 3 = 81$

4. 16
5. 32
6. 243
7. 216
8. 1000
9. 1

10.

Activity 4: Exponents and Calculators

1. 64
2. 32
3. 1728
4. 39.304
5. 6340
6. 16
7. 64
8. 10,000
9. 17.64
10. 1
11. 16
12. 8
13. 4
14. 2
15. 1
16. 625
17. 125
18. 25
19. 5
20. 1

$a° = 1$ (not zero!) $a \neq 0$

169

Activity 5: Algebraic Expressions

1. 8	12. 12
2. 9	13. 9
3. 24	14. 0
4. 10	15. 27
5. 1	16. 10
6. 10	17. 25
7. 8	18. 12
8. 14	19. 9
9. 13	20. 25
10. 14	21. 14
11. 5	22. 14

Space visitors message: TAKE ME TO YOUR ALGEBRA BOOK, EARTH PERSON.

Activity 6: Something of Interest

Quarter	Interest	Balance
3	$ 15.45	$1,045.68
4	15.69	1,061.37

Total Interest	Total Amount
$247.29	$3,247.29

Activity 7: The Compound Interest Formula

1. A = $13,727.86
2. I = $3,727.86
3. A = $6,734.28
4. A = $19,201.27
5. 36 periods, 9 years
6. $541.22
7. $1.00 was put into the formula for periods 1, 2, 3, and 4 or 1.02 y^x for x = 1, 2, 3, 4

Activity 8: Formula Review

1. 18	7. 90
2. 20	8. 64
3. 616	9. 1,540
4. 43.96	10. 462
5. 48	11. 38,808
6. 12.18	12. 18

A rocket ship.

Activity 9: Basic Properties

Step 1. 0+1+9+2+8+3+7+4+6+5

Step 2. 0+(1+9)+(2+8)+(3+7)+(4+6)+5

Step 3. (4×10)+5=45

2(a) 8×(40+3)=(8×40)+(8×3)=320+24=344

2(b) 6×(2+1/2)=(6×2)+(6×1/2)=12+3=15

   3. 1,500 – 30 = 1,470

Activity 10: Introduction to Equations

1. $y = 9$
2. $n = 19$
3. $y = 3$
4. $n + 9 - 9 = 12 - 9$ $n = 3$
5. $x + 15 - 15 = 23 - 15$ $x = 8$
6. $y - 8 + 8 = 15 + 8$ $y = 23$
7. $a - 23 + 23 = 56 + 23$ $a = 79$
8. $n = 28$
9. $y = 306$
10. $x = 83$
11. $m = 183$
12. $a = 76$
13. $b = 171$

Activity 11: An Equation Model

1. □ + ⦿⦿ = ⦿⦿⦿⦿

2. □ + □ = ⦿⦿⦿

3. □ + □ + ⦿⦿ = ⦿⦿⦿⦿⦿

Activity 12: Using a Model to Solve Equations (Addition/Subtraction)

1. $\square + $ ⚬⚬ = ⚬⚬⚬⚬

2. $\square + $ ⚬⚬⚬ = ⚬⚬⚬ ●

3. $\square + $ ⚬⚬⚬ = ⚬⚬⚬ ●●●●

4. $\square - $ (⚬⚬) = ●●● ●●● ●●●   (Note: These represent an alternate method.)

5. $\square - $ (⚬) = ●●● ●●●●

Activity 13: Solving Equations (Multiplication and Division)

1. $y = 4$

2. $y = 3$

3. $y = 3$

1. $\dfrac{\cancel{4}^{1}}{1} \times \dfrac{n}{\cancel{4}_{1}} = 3 \times 4 \qquad n = 12$

2. $\dfrac{\cancel{12}^{1}}{1} \times \dfrac{c}{\cancel{12}_{1}} = 24 \times 12 \qquad c = 288$

3. $\dfrac{\cancel{5}^{1}}{1} \times \dfrac{x}{\cancel{5}_{1}} = 15 \times 5 \qquad x = 75$

4. $\dfrac{5y}{5} = \dfrac{35}{5} \qquad y = 7$

5. $\dfrac{15n}{15} = \dfrac{225}{15} \qquad n = 15$

6. $\dfrac{18}{1} \times \dfrac{b}{18} = 34 \times 18 \qquad b = 612$

Bonus: $-\dfrac{\square}{\square}- = $ ●● ●● ●● $\quad y = 6$

Activity 14: Solving Equations with Whole Numbers (One Inverse Operation)

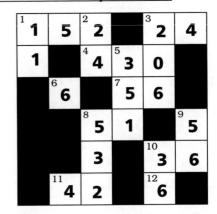

Activity 15: Solving Equations (Several Inverse Operations)

1. $y = 2$
2. $n = 3$
3. $x = 1$
4. $y = 1$
5. $y = 4$
6. $2y + 4 - 4 = 8 - 4$
   $\dfrac{2y}{2} = \dfrac{4}{2}$
   $y = 2$
7. $2n - 4 + 4 = 6 + 4$
   $\dfrac{2n}{2} = \dfrac{10}{2}$
   $n = 5$
8. $3x + 8 - 8 = 35 - 8$
   $\dfrac{3x}{3} = \dfrac{27}{3}$
   $x = 9$
9. $5n - 12 + 12 = 103 + 12$
   $\dfrac{5n}{5} = \dfrac{115}{5}$
   $n = 23$

10. $24a + 16 - 16 = 208 - 16$

$\dfrac{24a}{24} = \dfrac{192}{24}$

$a = 8$

11. $15c - 28 + 28 = 152 + 28$

$\dfrac{15c}{15} = \dfrac{180}{15}$

$c = 12$

Activity 16: Solving Equations with Whole Numbers (Several Inverse Operations)

1. 5	8. 3
2. 8	9. 12
3. 2	10. 1
4. 9	11. 13
5. 6	12. 11
6. 12	13. 20
7. 10	14. 5

Completed Dot-to-Dot Puzzle

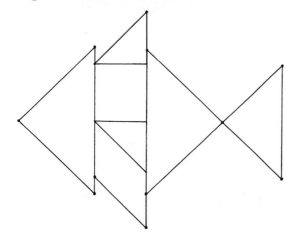

Assembled Puzzle Pieces

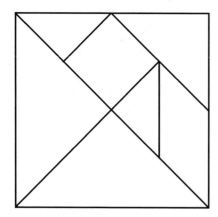

Activity 17: Solving Equations with Like Terms

1. $\square + \boxed{/} + \boxed{/} = \bullet\bullet\bullet\bullet\bullet$

$y = 5$

2. $\square \quad \bullet\bullet\bullet$

$\square = \bullet\bullet\bullet$

$\boxed{\square} \quad \bullet\bullet\bullet$

$n = 3$

3. $\boxed{/} + \dfrac{\boxed{/} + \square}{\boxed{/} + \boxed{\square}} = \dfrac{\bullet\bullet\bullet\bullet}{\bullet\bullet\bullet\bullet}$

$x = 4$

4. $\boxed{/}\boxed{/}\boxed{/}\boxed{/} = \bullet\bullet\bullet\bullet$

$a = 5$

5. $5y = 75$

$\dfrac{5y}{5} = \dfrac{75}{5} \qquad y = 15$

6. $12n = 144$

$\dfrac{12n}{12} = \dfrac{144}{12} \qquad n = 12$

7. $3c = 93$

$\dfrac{3c}{3} = \dfrac{93}{3} \qquad c = 31$

Activity 18: Solving Equations (Variables on Both Sides)

1.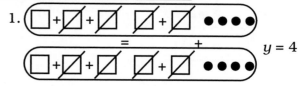

$y = 4$

2. $n = 6$

3. $y = 2$

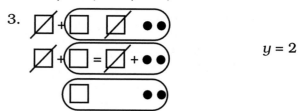

4. $3y - y = y - y + 12$

$\dfrac{2y}{2} = \dfrac{12}{2}$

$y = 6$

5. $5n - 2n = 2n - 2n + 12$

$\dfrac{3n}{3} = \dfrac{12}{3}$

$n = 4$

Activity 19: Equation Review

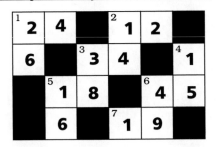

Activity 20: Introduction to Integers

BE POSITIVE.

Activity 21: Addition of Integers

1. 9
2. 3
3. –7
4. 0
5. 2
6. –2
7. –10
8. 0
9. –3
10. +5

Activity 22: Addition Patterns

1. 5
2. 7
3. 9
4. 6
5. –5
6. –7
7. –9
8. –6
9. –1
10. –1
11. –1
12. 0

13. Add as always and use the sign on the integers.
14. Find the difference between the integers and use the sign on the integer with the greater absolute value.
15. –17. Zero in each case.
18. –12
19. +8

20. –3
21. +4
22. –a

Activity 23: A Magic Circle (Addition of Integers)

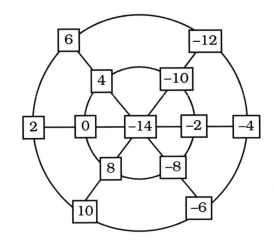

Activity 24: Addition of Integers

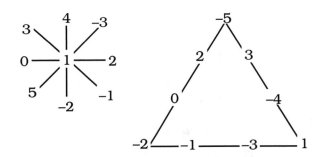

Activity 25: Subtraction of Integers

1. 5
2. –9
3. –2
4. –9
5. –3
6. 3
7. 0
8. –6
9. –5
10. –10

Activity 26: Subtraction Patterns

1. 5
2. –2
3. –10
4. –5
5. 10
6. 5
7. –2
8. –10
9. –5
10. 10

11. The problems on the right are adding the *opposite* of the number being subtracted in the problems on the left side.

12. 2

13. 14

14. –2

15. –14

16. –8

Activity 27: Addition/Subtraction of Integers

5	–9	1
–5	–1	3
–3	7	–7

Sum = –3

3	1	11
13	5	–3
–1	9	7

Sum = 15

9	–5	–4	6
–2	4	3	1
2	0	–1	5
–3	7	8	–6

Sum = 6

22	–6	–4	16
0	12	10	6
8	4	2	14
–2	18	20	–8

Sum = 28

Activity 28: Multiplication Patterns (Integers)

1. 12, 8, 4, 0, Decreasing
2. –4, –8, –12, Negative
3. –12, –9, –6, –3, 0, Increasing
4. 3, 6, 9

X	–3	–2	–1	0	+1	+2	+3
+3	–9	–6	–3	0	+3	+6	+9
+2	–6	–4	–2	0	+2	+4	+6
+1	–3	–2	–1	0	+1	+2	+3
0	0	0	0	0	0	0	0
–1	3	2	1	0	–1	–2	–3
–2	6	4	2	0	–2	–4	–6
–3	9	6	3	0	–3	–6	–9

1. Positive
2. Negative
3. Negative
4. Positive

In multiplication, if the signs are different the answer is (negative) and if the signs are the same the answer is (positive.)

Activity 29: A Division Pattern

1. 3   2. –3   3. –3   4. 3
5. (a) Positive   (b) Negative
6. 0   7. You can't divide by zero.

Activity 30: Working with Rational Numbers

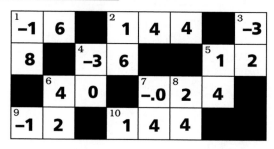

Activity 31: Rational Number Review

1. 5
2. −8
3. −5
4. −9
5. 0
6. −1.9
7. $\frac{1}{12}$
8. −6
9. −16
10. 15
11. 7
12. −.67
13. $1\frac{1}{6}$

14. 10
15. 12
16. −20
17. −24
18. .06
19. 9
20. $-\frac{1}{2}$
21. −3
22. 4
23. 1
24. $-\frac{8}{9}$
25. 5

Dot-to-Dot Puzzle

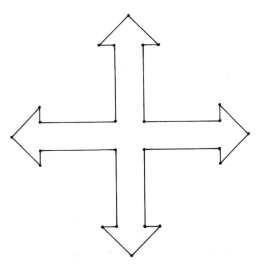

They are sometimes called *directed* numbers.

Activity 32: Exponents and Integers

1. 32, 16, 8, 4
2. $\frac{1}{2}$
3. 2, 1, $\frac{1}{2}$
4. 625, 125, 25, 5, 1, $\frac{1}{5}$, $\frac{1}{25}$
5. 27, 9, 3, 1, $\frac{1}{3}$, 2, 3

6. $\frac{1}{2^3}$ 7. $\frac{1}{2^3}$ 8. $\frac{1}{4^2}$ 9. $\frac{1}{6^3}$ 10. $\frac{1}{9^3}$

11. a, 1, $\frac{1}{a^n}$

Activity 33: An Exponent Cartoon (Zero and Negative Exponents)

1. 1
2. $\frac{1}{9}$
3. $\frac{1}{16}$
4. $-\frac{1}{8}$
5. 16
6. $\frac{3}{a^2}$

7. 5
8. $\frac{1}{3}$
9. a^4
10. $\frac{1}{a^2}$
11. 1

Puzzle

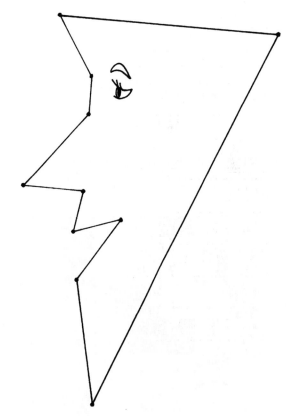

Activity 34: Equations with Rational Numbers (One Inverse Operation)

¹1	7	²6		³6	0
9		⁴3	⁵2	0	
	⁶±6		⁷2	4	
		⁸2	5		⁹2
	2			¹⁰−4	4
	¹¹1.	4		¹²2	

Activity 35: Equations with Rational Numbers (Several Inverse Operations)

7	27	29	1
17	13	11	23
9	21	19	15
31	3	5	25

Sum = 64

Activity 36: An Equation Code (Like Terms)

WORK WITH LIKE TERMS IS EASY.

Activity 37: Solving Equations (Variables on Both Sides of the Equal Sign)

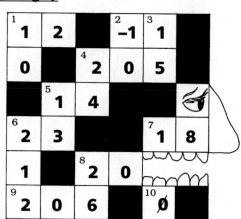

¹1	2		²−1	³1	
0		⁴2	0	5	
	⁵1	4			
⁶2	3			⁷1	8
1		⁸2	0		
⁹2	0	6		¹⁰∅	

Activity 38: Solving Inequalities

1. 8
2. $x > 4$
3. 6
4. $y < 5$
5. −4
6. $n < −7$
7. 7
8. $y < 3$
9. 2
10. 3
11. 9
12. $y > 7$
13. −6
14. −3
15. −2
16. $y > −3$
17. −5
18. .8
19. −27
20. ∅
21. $y < 6$
22. 10
23. $y < 9$
24. 1.2

Puzzle

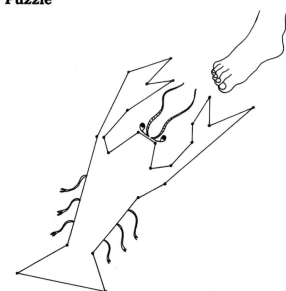

Activity 39: Addition of Polynomials

GOOGOLPLEX = $10^{10^{100}}$

Activity 40: Addition and Subtraction of Polynomials

1. $3y^2 + 11$
2. $3x + 2$
3. $5x − 2$
4. $4y^2 + 3y$

Dot-to-Dot Puzzle

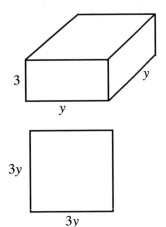

1. $5y + 6$
2. $6y + 6$
3. $2x^2 - 1$
4. $3y^2$
5. $5a + 3b + 5$
6. 0
7. $4a^2 + a - 2$
8. $3a^3 + 6a + 2$
9. $8xy$
10. $5z^3 - z^2 - 3z$
11. $7a^2$
12. $4y^3 - 5y^2$
13. $a^3 + 5ab$
14. $2a + 2$
15. $2y + 2x$
16. $-2b$
17. $-2x - 3$
18. $2xy$
19. -2
20. 4
21. $8x$
22. $5a^2$
23. $-x + 3$
24. 6

Activity 41: Multiplication of Monomials

2. $a \cdot b = ab$
3. $4 \cdot 4 = 4^2 = 16$
4. $y \cdot y = y^2$
5. $2c \cdot 4d = 8cd$
6. $x \cdot y \cdot z = xyz$
7. $y \cdot y \cdot y = y^3$
8. $2x \cdot 3y \cdot 4z = 24xyz$
9. 4^2 can be represented as a 4 × 4 square, 5^3 as a 5 × 5 × 5 cube.
10.

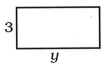

Activity 42: Advanced Exponent Patterns

1. 5, 4, 6, 5
2. add
3. a^{n+m}
4. no
5. 2
6. 3
7. a^{n-m}
8. 1
9. 0
10. 5
11. 5
12. 9
13. 1
14. 12
15. $8y^3$
16. $27x^3$
17. a^6b^9
18. $x^4y^2z^6$
19. $c^{12}d^4$
20. $(ab)^x = a^x b^x$
21. 3^8
22. y^6
23. 2^{10}
24. 8^4
25. x^9
26. $(a^x)^y = a^{xy}$

Activity 43: An Exponent Cross-Number Puzzle

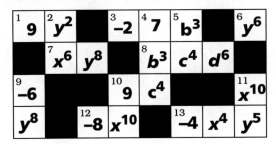

Activity 44: Multiplying a Polynomial by a Monomial

1. $3y$ ⎡$3y^2$⎤ $+ \; 3y$ ⎡$12y$⎤ $=$ ⎡$3y^2$ | $+ 12y$⎤ $\quad 3y$
   $\quad\quad\;\; y \quad\quad\quad\quad\; 4 \quad\quad\quad\quad\; y \quad\;\; + 4$

2. $4x$ ⎡$4xy$⎤ $+ \; 4x$ ⎡$24x$⎤ $=$ ⎡$4xy$ | $+ 24x$⎤ $\quad 4x$
   $\quad\quad\;\; y \quad\quad\quad\quad\; 6 \quad\quad\quad\quad\; y \quad\;\; + 6$

3. $3a(c + 4) = 3a$ ⎡$3ac$⎤ $+ \; 3a$ ⎡$12a$⎤ $=$ ⎡$3ac \;\; + 12a$⎤ $\quad 3a$
   $\quad\quad\quad\quad\quad\quad\quad\;\; c \quad\quad\quad\quad\;\; 4 \quad\quad\quad\quad c \quad\; + 4$

4. $-5a^2 \cdot a^3 + (-5a^2) \cdot 2a = -5a^5 + (-10a^3)$

Activity 45: A Polynomial Code (Monomial × Polynomial)

WHAT PROPERTY MAKES THIS WORK?
(DISTRIBUTIVE)

Activity 46: Multiplying Binomials

1. 30
2. $x^2 + 7x + 12$
3. $2x^2 + 7x + 6$
4. 14
5. $y^2 - y - 6$
6. $6a^2 - 8a - 8$
7. 32
8. $a^2 - b^2$
9.

	4	+ 2
4	16	8
+		
3	12	6

$= 16 + 8 + 12 + 6 = 42$

10.

$x + 2$

	x²	2x
x + 3	3x	6

$= x^2 + 2x + 3x + 6$
$= x^2 + 5x + 6$

Activity 47: (a + b) (a – b)— Multiplication of Binomials (A Special Pattern)

1. $x^2 - 4$
2. $4y^2 - 9$
3. $y^2 - 9$
4. $9a^2 - 4$
5. $x^2 - y^2$

6. The binomials are the same, except for the sign.
7. Each answer is the difference of 2 perfect squares.
8. $x^2 - 9$
9. $n^2 - 4$
10. $4y^2 - 16$
11. 1596
12. 391
13. 884

Activity 48: (a + b)², (a – b)²

1. $25 + 15 + 15 + 9 = 64$
2. $y^2 + 3y + 3y + 9 = y^2 + 6y + 9$
3. $(2x + 4)(2x + 4) = 4x^2 + 8x + 8x + 16 = 4x^2 + 16x + 16$
4. $(a + b)(a + b) = a^2 + ab + ab + b^2 = a^2 + 2ab + b^2$
5. $a^2 + 8a + 16$
6. $x^2 - 6x + 9$
7. $4y^2 + 20y + 25$
8. $4y^2 - 12y + 9$
9. $x^2 + 2xy + y^2$
10. $x^2 - 2xy + y^2$
11.

	5	+ 2
5	25	10
+		
2	10	4

$= 49$

12.

	a	+ b
a	a²	ab
+		
b	ab	b²

$= a^2 + 2ab + b^2$

13.
$$\begin{array}{c} 3 + 5 \\ \begin{array}{c}3\\+\\5\end{array}\begin{array}{|c|c|}\hline 9 & 15 \\\hline 15 & 25 \\\hline\end{array} = 64 \end{array}$$

14.
$$\begin{array}{c} 2y + 3 \\ \begin{array}{c}2y\\+\\3\end{array}\begin{array}{|c|c|}\hline 4y^2 & 6y \\\hline 6y & 9 \\\hline\end{array} = 4y^2 + 12y + 9 \end{array}$$

Activity 49: (a + b)³—An Extra Project

Have students show you their models.

Activity 50: Polynomial × Polynomial (Magic Square)

16	2	3	13
5	11	10	8
9	7	6	12
4	14	15	1

The sum of each row is 34.

Activity 51: Dividing a Polynomial by a Monomial

1. $2y + 3$
2. $x + 3$
3. $4y + 3$
4. $3y^2 + 4y + 2$
5. $2y + 1$
6. 8
7. $a + 4$
8. $y + 5$
9. $2n + 4$
10. $a^2 + 2a + 5$
11. $4n - 3$
12. $4x - 3$
13. $4y - 1$
14. $y - x$
15. $b^2 - 4$
16. $2a + \dfrac{5}{3}$
17. $3a - \dfrac{b}{a}$
18. $3 - \dfrac{5}{y}$
19. $\dfrac{5}{2} - 3y$
20. $4a - 3$

Activity 52: Dividing a Polynomial by a Binomial

1. $n + 4$

	$4n$
$3n$	

2. $n + 2$

n^2	$2n$
$5n$	10

Activity 53: Division of Polynomials

1. 4
2. y^2
3. $-y^3$
4, $5y^2$
5. $-4x^2$
6. $2x$
7. $\dfrac{1}{3x}$
8. $\dfrac{2x}{y}$
9. $\dfrac{-x^3}{3y^2}$
10. $x + 2$
11. $x - 3$
12. $x + 3$
13. $x^2 + 2x + 3$
14. $2x + 6$
15. $x - 4$
16. 4

Activity 54: Finding the GCF

¹**1**	²**2**		³**1**	⁴**3**	
	⁵**4**	**x**		**a**	
⁶**4**			⁷**a²**	**b**	
⁸**a³**	⁹**b⁴**		**b⁴**		
	¹⁰**d²**	**a⁴**	**c²**		
¹¹**6**				¹²**1**	
x²		¹³**9**	¹⁴**b**		
	¹⁵**3**		¹⁶**c²**	¹⁷**d**	
	¹⁸**2**	**7**	**d³**	**e²**	

3. $5y^2 (y + 3)$
4. $4ab^2 (ab + 2)$
5. $5 (a^2 + a - 3)$
6. $4a (a^2 + 3a + 4)$
7. $4xy^2 (6x^2 - 5x + 4)$
8. $(y^2 + 3) (y + 2)$

Activity 55: Monomial Factors

1. $3 (y^2 - 3)$
2. $3y (4y^2 + 1)$

Activity 56: A Polynomial Pattern (Factoring $x^2 + bx + c$)

$(x + r) (x + s)$	$x^2 + bx + c$	b	c	$r + s$	$r \times s$
$(x + 3) (x + 2) =$	$x^2 + 5x + 6$	5	6	$3 + 2 = 5$	$3 \times 2 = 6$
$(x + 5) (x + 3) =$	$x^2 + 8x + 15$	8	15	$5 + 3 = 8$	$5 \times 3 = 15$
$(x + 4) (x - 2) =$	$x^2 + 2x - 8$	2	-8	$4 + (-2) = 2$	$4 \times (-2) = -8$
$(x - 3) (x - 2) =$	$x^2 - 5x + 6$	-5	$+6$	$-3 + (-2) = -5$	$-3 \times (-2) = +6$
$(x - 3) (x + 2) =$	$x^2 - x - 6$	-1	-6	$-3 + 2 = -1$	$-3 \times (+2) = -6$

1. They are the same.
2. They are the same.

Activity 57: Factoring $x^2 + bx + c$)

1. $(x + 2) (x + 3)$
2. $(x + 1) (x - 2)$
3. $(x + 3)^2$
4. $(x + 5) (x - 4)$
5. $(x + 2) (x - 3)$
6. $(x - 2) (x - 3)$
7. $(x + 3) (x + 4)$
8. $(x + 4) (x + 2)$
9. $(x - 3)^2$
10. $(x + 8) (x - 4)$
11. $(x - 4)^2$
12. $(x + 9) (x + 4)$
13. $(x - 6)^2$
14. $(x + 6)^2$
15. $(x + 9) (x - 4)$
16. $(x - 6) (x + 1)$
17. $(x - 1)^2$
18. $(x + 15) (x - 1)$
19. $(x - 3) (x - 4)$
20. $(x + 1)^2$

Activity 58: Factoring $ax^2 + bx + c$

6	7	2
1	5	9
8	3	4

Sum = 15

Activity 59: An Equation Code (Solving Quadratic Equations by Factoring)

THIS IS TOO HARD FOR ME.

Activity 60: Rational Expressions

1. 0   2. –4   3. –3 or 2

4	14	15	1
9	7	6	12
5	11	10	8
16	2	3	13

Sum = 34

Activity 61: Simplifying Rational Expressions

1. y

2. $\dfrac{x}{y}$

3. $\dfrac{3x}{4y}$

4. y^4

5. $\dfrac{y}{3x}$

6. $\dfrac{-x^2}{5}$

7. $\dfrac{y}{x+y}$

8. y^4

9. $x - y$

10. $\dfrac{x+2}{x-2}$

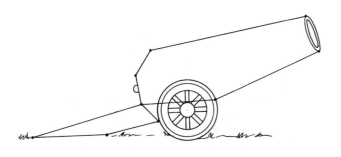

Activity 62: Multiplication and Division of Fractions

YOU ARE THE WINNER.

Activity 63: Outdoor Math (Similar Triangles)

h = 18 feet, answers will vary.

Activity 64: Percent Equations (Review)

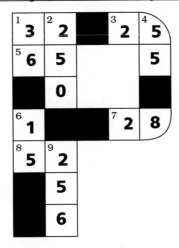

Activity 65: Addition and Subtraction of Like Fractions

4	9	2
3	5	7
8	1	6

Sum = 15

Activity 66: Addition and Subtraction of Fractions with Different Denominators

1. $\dfrac{4a + 15c}{6}$

2. $\dfrac{7y^2}{12}$

3. $\dfrac{12 + y}{3y}$

4. $\dfrac{5y^2}{4}$

5. $\dfrac{17}{3y}$

6. $\dfrac{y + 6}{10}$

7. $\dfrac{7 - y}{12}$

8. $\dfrac{3y + 4}{6}$

9. $\dfrac{17y + 9}{6}$

10. $\dfrac{5y - 3}{y^2}$

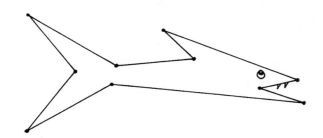

Activity 67: Ordered Pairs

1. (–2, 3)
2. (3, 4)
3. (–4, –4)
4. From the origin go left 3 and up 2.
5. From the origin go right 3 and down 4.

(Diagram)

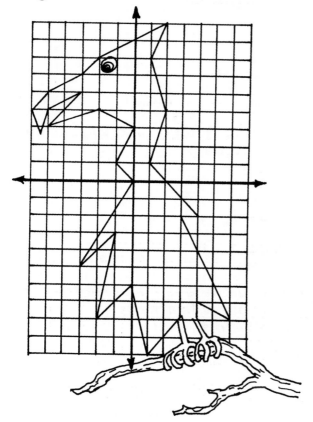

Activity 68: Graphing Equations

Example 2: Graph $y = 2x$

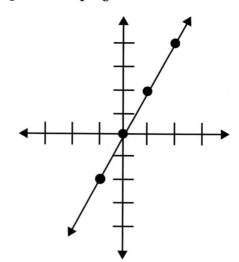

1.

x	y
0	4
–2	0
1	6
2	8
–1	2

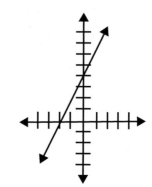

2.

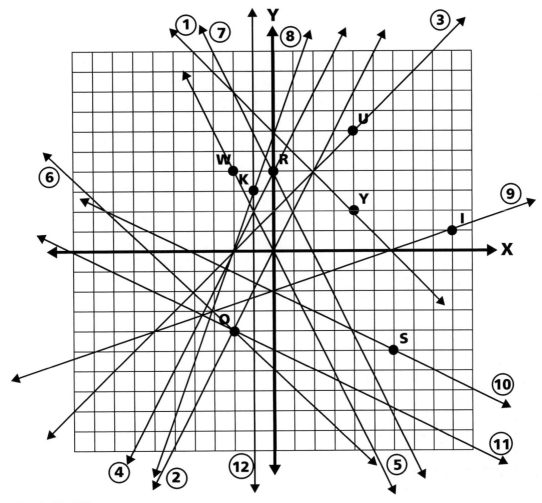

YOUR WORK IS OK.

Activity 69: Solving Systems of Equations Graphically

1. (3, 5)
2. (3, 2)
3. (4, –1)
4. (1, –3)
5. (3, –4)
6. (–3, –4)
7. (–3, 4)
8. (3, 5)

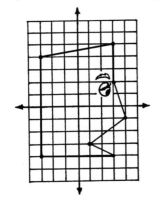

Activity 70: Solving Systems of Equations (Addition and Multiplication Method)

2	5	6
4	0	3
2	8	4

1	3	6
4	2	3
3	8	0

1	5	2
3	2	4
2	6	6

Activity 71: Solving Systems of Equations (Substitution Method)

[1]2	0		[2]1	2
4		[3]3	0	
				[4]1
[5]1	[6]8		[7]2	4
	2		3	

Activity 72: Finding the Slope of a Line

1. (1, –2), (–3, 4)

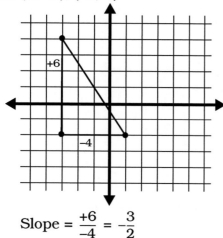

$$\text{Slope} = \frac{+6}{-4} = -\frac{3}{2}$$

2. (3, 2), (4, –2)

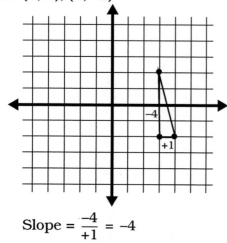

$$\text{Slope} = \frac{-4}{+1} = -4$$

3. (1,1), (–3, –2)

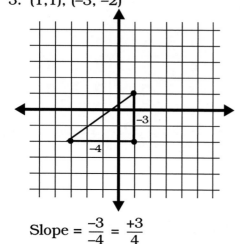

$$\text{Slope} = \frac{-3}{-4} = \frac{+3}{4}$$

4. – 3/2
5. –4
6. + 3/4

7. 0
8. Vertical Line (No Slope)

Activity 73: Square Root (Perfect Squares)

1. 6
2. 7
3. 10

4. 9
5. 12
6. 1

Square Roots Puzzle

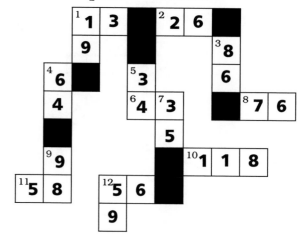

Activity 74: Square Root (Irrational Numbers)

1. 3.6
2. 4.6
3. 9.3
4. 11.4
5. 9.9
6. 11.6

7. 12.2
8. 15.4
9. 18.1
10. 4.4
11. 5.7
12. 1.9

$\pi = 3.14$

Activity 75: An Introduction to the Pythagorean Theorem

Length of sides			a^2	b^2	$a^2 + b^2$	c^2	Is a right triangle formed?
a	b	c					
3	4	5	9	16	9 + 16 = 25	25	yes
5	12	13	25	144	169	169	yes
4	5	7	16	25	41	49	no
6	8	10	36	64	100	100	yes
12	16	20	144	256	400	400	yes
6	10	13	36	100	136	169	no

$a^2 + b^2 = c^2$

1. $c = 13$, $a = 3$

Activity 76: Square Root Approximation (Pythagorean Theorem)

2. 5.8
3. 3.2
4. 3.6
5. 4.5
6. 4.2
7. 6.4

Activity 77: A Pattern for Completing the Square

1. ±7   2. 10, –2   3. 1, –7

1	8	16	4	$4^2 = 16$	$(x + 4)(x + 4)$ or $(x + 4)^2$
1	–6	9	–3	$(-3)^2 = 9$	$(n - 3)(n - 3)$ or $(n - 3)^2$
1	2	1	1	$1^2 = 1$	$(y + 1)(y + 1)$ or $(y + 1)^2$

4. a = one   5. c 6. They are perfect squares.

Activity 78: The Quadratic Formula

1. $(-3, -2)$

2. $(2, -3)$

3. $(-\frac{1}{2}, -3)$

4. $(3, -5)$

5. $(5, 2)$

6. $(\frac{2}{3}, -1)$

7. $(4, 3)$

8. $(\frac{1}{2}, 3)$

9. $(-\frac{3}{2}, 2)$

10. $(-\frac{2}{3}, 2)$

11. -3

12. $(5, -2)$

13. $(\frac{1}{2}, 4)$

14. $(-\frac{1}{2}, -4)$

15. 3

16. $\frac{3 \pm \sqrt{5}}{2}$

17. $(0, -3)$

18. ± 1

19. $(0, 3)$

20. -5

21. $\frac{1 \pm \sqrt{13}}{2}$

22. $\frac{1 \pm \sqrt{5}}{2}$

23. ± 4

24. $2 \pm \sqrt{3}$

Activity 79: A Pythagorean Experiment

Discuss the logic of this activity with your students.

Activity 80: Finding the Distance Between Two Points on a Graph

1. 10   2. 5.66   3. 3.61

4. 10   5. 5.66   6. 3.61

7. $x_2 - x_1$, gives the horizontal leg

8. $y_2 - y_1$, gives the vertical leg

9. $(x_2 - x_1)^2 + (y_2 - y_1)^2 = c^2 (a^2 + b^2 = c^2)$

10. $\sqrt{c^2} = c =$ the distance

J. WESTON
WALCH
PUBLISHER

Share Your Bright Ideas with Us!

We want to hear from you! Your valuable comments and suggestions will help us meet your current and future classroom needs.

Your name_____Date_____

School name_____Phone_____

School address_____

Grade level taught_____Subject area(s) taught_____Average class size_____

Where did you purchase this publication?_____

Was your salesperson knowledgeable about this product?   Yes_____   No_____

What monies were used to purchase this product?

_____School supplemental budget   _____Federal/state funding   _____Personal

Please "grade" this Walch publication according to the following criteria:

Quality of service you received when purchasing ....A   B   C   D   F
Ease of use...A   B   C   D   F
Quality of content..A   B   C   D   F
Page layout ....A   B   C   D   F
Organization of material ....A   B   C   D   F
Suitability for grade level ....A   B   C   D   F
Instructional value..A   B   C   D   F

COMMENTS:_____

_____

What specific supplemental materials would help you meet your current—or future—instructional needs?

_____

Have you used other Walch publications? If so, which ones?_____

May we use your comments in upcoming communications?   _____Yes   _____No

Please **FAX** this completed form to **207-772-3105**, or mail it to:

Product Development, J. Weston Walch, Publisher, P.O. Box 658, Portland, ME 04104-0658

We will send you a **FREE GIFT** as our way of thanking you for your feedback.   **THANK YOU!**

12 + 12 + 12

12 × 7 = 72

$4\overline{)72}$
18
4

32